A Guide to Sampling Freshwater Mussel Populations

Financial support for the printing of this monograph provided by the

Freshwater Mollusk Conservation Society

A Guide to Sampling Freshwater Mussel Populations

David L. Strayer
Institute of Ecosystem Studies, Post Office Box AB, Millbrook, New York 12545, USA

David R. Smith
U.S. Geological Survey, Leetown Science Center, Kearneysville, West Virginia 25430, USA

American Fisheries Society Monograph 8

American Fisheries Society
Bethesda, Maryland
2003

A suggested citation format for this monograph is

Strayer, D. L., and D. R. Smith. A guide to sampling freshwater mussel populations. American Fisheries Society, Monograph 8, Bethesda, Maryland.

Cover photographs courtesy of Chris Barnhart, Southwest Missouri State University (bottom left photo), and the Leetown Science Center, USGS.

Library of Congress Control Number 2003104869

ISBN 1-888569-50-6 ISSN 0362-1715

Printed in the United States of America on acid-free paper

American Fisheries Society
5410 Grosvenor Lane, Suite 110
Bethesda, Maryland 20814-2199
USA

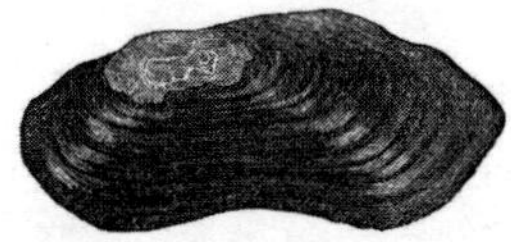

Contents

Preface

We wrote this guide to help field biologists conduct better surveys of freshwater mussel populations. Spurred largely by acute conservation concerns, biologists from government agencies, consulting firms, and academics now devote considerable effort to assessing the size, distribution, status, and trends of freshwater mussel populations. It is our impression that these studies do not always use the best possible designs and methods, which is a pity, because a suboptimal design can needlessly weaken otherwise good efforts.

Little in the guide is new; indeed, most of the material we discuss is available in books on statistics and sampling designs. Nevertheless, it is our experience that mussel biologists are not using these general texts effectively, perhaps because the material they seek is embedded in a large amount of highly technical material that is not directly relevant to them. Furthermore, freshwater mussels have several traits that help to dictate good survey design—their populations often are sparse or patchy; the animals are large and readily identifiable in the field, cryptic, and not very motile; and their habitats are not always readily accessible to air-breathing terrestrial surveyors. We have focused on sampling designs and methods that take into account these special traits and have tried to write in simple, nontechnical language wherever possible.

The guide opens with a discussion of several considerations and a question that should direct the design of any study of mussel populations. We then present sampling designs and methods that may be useful to mussel biologists. We also discuss examples of study designs that address several common objectives of studies of mussel populations. We hope that field biologists will find much of the information they need to conduct good studies of freshwater mussel populations in this guide.

We appreciate the encouragement and financial support from Kevin Cummings, Heidi Dunn, and the board and membership of the Freshwater Mollusk Conservation Society, who helped defray the cost of publication. We thank Bill Lellis for letting us use his beautiful survey data from the Delaware River, Mary Christman and three anonymous reviewers for their careful and constructive reviews, and John Van Hassel for coordinating the review process. D.R.S. gratefully acknowledges his colleagues Rita Villella and David Lemarié, who have worked with him step-by-step to evaluate sampling techniques. Penelope Pooler assisted with modeling of distance sampling data. This is a contribution to the program of the Institute of Ecosystem Studies.

Symbols and Abbreviations

The following symbols and abbreviations may be found in this monograph without definition.

cm	centimeter
d	day
e.g.	(exempli gratia) for example
h	hour
i.e.	(id est) that is
km	kilometer
m	meter
min	minute
mm	millimeter
no.	number
s	second
α	probability of type I error
%	per cent (per hundred)

Abstract

Freshwater mussels are the most imperiled group of organisms in North America, play important roles in freshwater ecosystems, are useful indicators of water quality, and support regionally important commercial fisheries. Hence, scientists and managers often need reliable information about freshwater mussel populations. Currently, this information is gathered using a wide variety of sampling designs and methods, not all of which are appropriate or ideal. We critically review sampling designs and methods that might be useful for sampling freshwater mussel populations and offer exemplary designs to meet several common objectives.

The choice of appropriate sampling design and method depends on five central considerations: the goals of the survey, the definition of the target population, the resources of the investigator, the characteristics of the study site, and prior knowledge about the target population. Surveys should be designed only after carefully considering these five issues. After a survey is designed, but before it is implemented, the investigator must consider rigorously whether the anticipated results of the survey will satisfy its goals.

We review the use of complete censuses, informal sampling, simple random sampling, systematic sampling, stratified sampling, double sampling, adaptive sampling, two-stage designs, regression designs, mark–recapture methods, and distance sampling. All of these designs are useful for assessing mussel populations in defined circumstances, but most have important caveats that limit their use. Within any of these designs, small quadrats (e.g., 0.25 m^2) often are useful sampling units for mussels, but strip transects or larger quadrats also are useful, especially in cases where sampling units are difficult or expensive to set up. Mussels may be sampled using visual searches, tactile searches, sediment collection (by hand or grabs) and sieving, or special methods such as brail bars, dredges, or muskrat middens. As is the case with sampling designs, each of these sampling methods is useful in certain defined situations. We describe how the precision or power of a proposed survey can be estimated to evaluate the likelihood that a survey will meet its goals.

Finally, we offer suggestions of useful survey designs and methods to meet several common objectives of mussel surveys, including determining the presence of a species in a small area or in a large drainage basin, estimating population size or density, comparing mussel populations at two or more sites (including cases where one site is affected by human impacts), and assessing temporal change at a site or throughout a large drainage basin.

Introduction

Freshwater mussel populations are increasingly frequent objects of study. Many species of mussels have been endangered or eliminated by human activities (Bogan 1993; Williams et al. 1993; Master et al. 2000), and they may play important roles in freshwater ecosystems (Welker and Walz 1998; Strayer et al. 1999; Vaughn and Hakenkamp 2001), be useful monitors of water pollution (e.g., Green et al. 1989; Metcalfe and Charlton 1990; Elder and Collins 1991; Naimo 1995), and support regionally important commercial fisheries (Carlander 1954; Bowen et al. 1994; Claassen 1994; Ziuganov et al. 1994). As a result, biologists often need to estimate presence, extent, density, or size of freshwater mussel populations to evaluate the status of rare species (e.g., Strayer et al. 1996; Rogers et al. 2001), judge whether a population is declining (e.g., Strayer and Smith 1996), assess the influence of a human activity (e.g., Smith et al. 2001), quantify the potential effects of mussels on other parts of the ecosystem (e.g., Nalepa et al. 1991; Strayer et al. 1994), or pursue specific research on mussel biology or ecology (e.g., Downing et al. 1989). Many kinds of sampling gear and sampling designs have been used to evaluate freshwater mussel populations. These options differ widely in cost, ease of use, and suitability to address various objectives. The purpose of this guide is to review designs and gear for sampling populations of freshwater mussels and to help biologists design a sampling program that is well suited to their objectives. The paper consists of three main parts: a review of sampling designs, a review of sampling gear, and a discussion of approaches that we recommend in order to meet some common objectives. First, though, we briefly discuss five central considerations and an important question that must be taken into account when designing an effective sampling program.

Five Guiding Considerations and an Important Question

The design of an appropriate sampling program begins with five overriding considerations (Figure 1A). These considerations should be kept in mind throughout the planning process.

(1) Goals of the survey. Before you begin a mussel survey, you need to think clearly about its purpose. Why is the survey being done? Is it to determine which species live at a site? To determine population density or size for one or more

and the goal was to estimate density of freshwater mussels throughout the river. However, a large portion of the river was inaccessible, and it would be unwise to extrapolate observations made at bridge sites to the whole river, so the target population must be reduced from the "whole river" to "that portion of the river near bridge sites." Also, the survey objective must be reduced from estimating density throughout the river to estimating density near bridge sites.

(3) Resources of the investigator. How much time and labor can be devoted to the survey? Are special resources (e.g., scuba divers, taxonomically trained surveyors, boats) available? Survey goals must be realistically matched with available resources. If resources are insufficient to meet the goal of the survey, then additional resources must be mustered (and a carefully thought-out survey plan may be useful in justifying a request for additional resources) or the goals of the study must be scaled back.

(4) Characteristics of the study site. Does the study site or target species have special characteristics that will affect the performance of the survey? Such characteristics may affect the suitability and cost of various sampling devices and, therefore, dictate the sampling design. For instance, visual searches cannot be done in highly turbid waters, PONAR grabs are useless in cobble substrates, and scuba diving may be too costly or unsafe in deep, swift waters. Likewise, the size and habits of a mussel species will affect the suitability of visual searches for surveys and the amount of excavation required to estimate density.

(5) Prior knowledge about the mussel populations. Often an investigator will know something about the study population before a formal survey is begun. For example, you may know (or suspect) that mussels are found chiefly near the banks or that large mussel beds occur at known sites in the study area. (However, be careful about assuming habitat associations that have not been verified; mussels are often found in unexpected places and, conversely, not found where expected.) Prior information is immensely valuable in designing surveys because it can be used to set up stratified or adaptive designs, which can be much more powerful than purely random designs. In fact, it often will be useful to conduct preliminary, informal surveys to gather information before designing the formal survey.

An important question. Once a survey design is sketched out, a final question should be answered to assess whether the survey plan is adequate: "Will the results of the survey meet its goals?" (Figure 1B). In particular, will the survey results be precise enough to provide useful information toward the goals? Because most mussel populations are so patchy, data collected in mussel surveys usually have a large variance. This variance can be so large that the survey data are essentially useless in addressing the survey goals.

Two important steps can reduce the likelihood that this unhappy situation will develop. First, the survey goals should be stated in quantitative terms. For example, instead of simply saying that you are interested in detecting the temporal change in the brook floater *Alasmidonta varicosa* population in the Neversink River, New York, think hard about how large of a change you want to detect. Are you monitoring to detect subtle changes in population size (say ±10% between surveys), or will

Introduction

Freshwater mussel populations are increasingly frequent objects of study. Many species of mussels have been endangered or eliminated by human activities (Bogan 1993; Williams et al. 1993; Master et al. 2000), and they may play important roles in freshwater ecosystems (Welker and Walz 1998; Strayer et al. 1999; Vaughn and Hakenkamp 2001), be useful monitors of water pollution (e.g., Green et al. 1989; Metcalfe and Charlton 1990; Elder and Collins 1991; Naimo 1995), and support regionally important commercial fisheries (Carlander 1954; Bowen et al. 1994; Claassen 1994; Ziuganov et al. 1994). As a result, biologists often need to estimate presence, extent, density, or size of freshwater mussel populations to evaluate the status of rare species (e.g., Strayer et al. 1996; Rogers et al. 2001), judge whether a population is declining (e.g., Strayer and Smith 1996), assess the influence of a human activity (e.g., Smith et al. 2001), quantify the potential effects of mussels on other parts of the ecosystem (e.g., Nalepa et al. 1991; Strayer et al. 1994), or pursue specific research on mussel biology or ecology (e.g., Downing et al. 1989). Many kinds of sampling gear and sampling designs have been used to evaluate freshwater mussel populations. These options differ widely in cost, ease of use, and suitability to address various objectives. The purpose of this guide is to review designs and gear for sampling populations of freshwater mussels and to help biologists design a sampling program that is well suited to their objectives. The paper consists of three main parts: a review of sampling designs, a review of sampling gear, and a discussion of approaches that we recommend in order to meet some common objectives. First, though, we briefly discuss five central considerations and an important question that must be taken into account when designing an effective sampling program.

Five Guiding Considerations and an Important Question

The design of an appropriate sampling program begins with five overriding considerations (Figure 1A). These considerations should be kept in mind throughout the planning process.

(1) Goals of the survey. Before you begin a mussel survey, you need to think clearly about its purpose. Why is the survey being done? Is it to determine which species live at a site? To determine population density or size for one or more

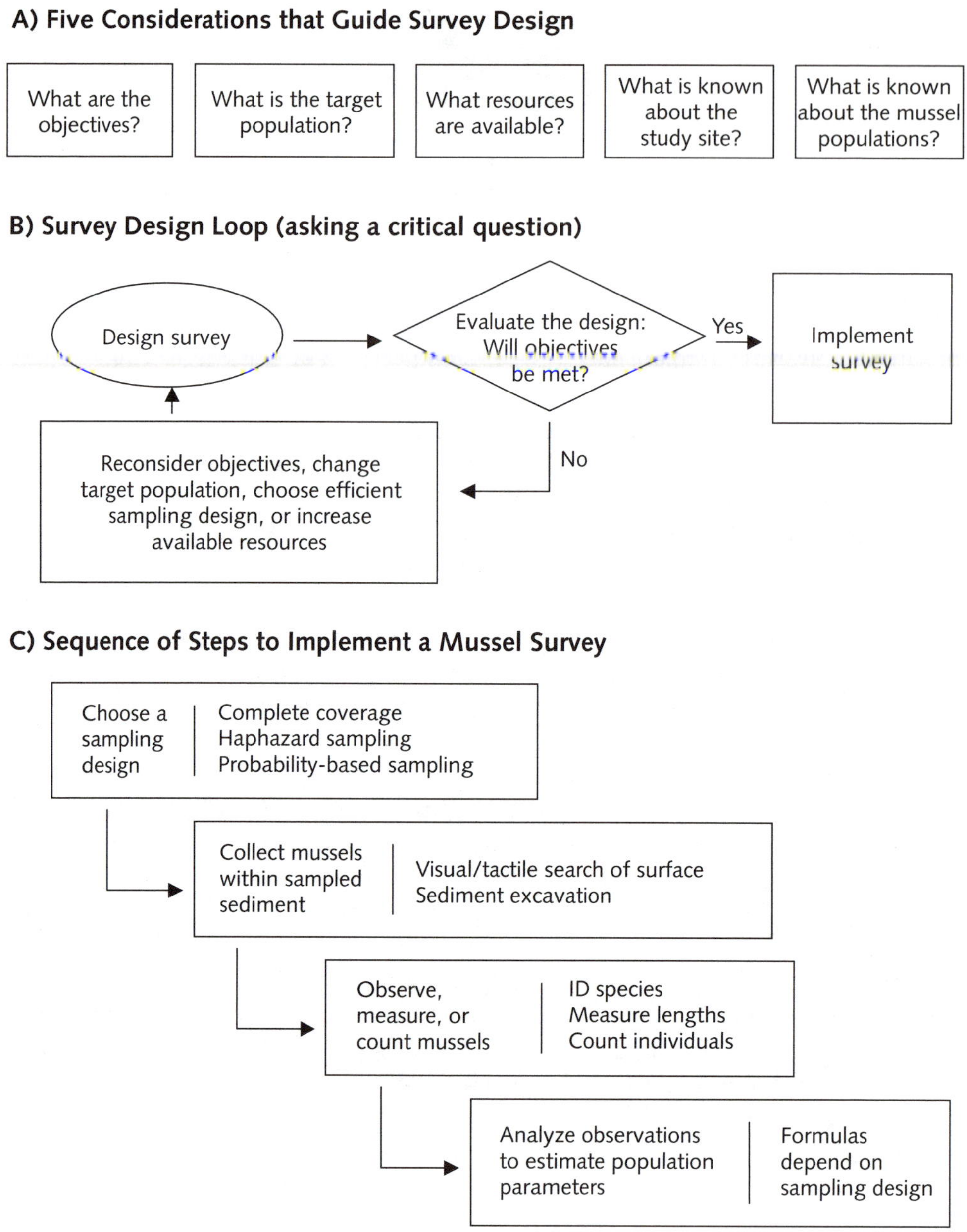

FIGURE 1. Diagram showing (A) the questions that guide survey design, (B) the iterative nature of survey planning where initial design is followed by asking whether the survey design is sufficient to meet survey objectives, and (C) the hierarchical nature of survey implementation where sampling of a physical area by a chosen sampling design is followed by collection of mussels within the sampled area or sediment, which is followed by observations of collected mussels, which is followed by analysis of the observations.

species? To estimate attributes (e.g., body size, sex ratio) of one of these populations? To track changes in mussel populations over time, including changes in population size or rates of survival, recruitment, or movement? To compare mussel populations at two or more sites, perhaps as an assessment of suspected impacts at one or more of the sites? To describe the spatial distribution of mussels at a site? Each of these goals is best addressed using a particular design, and only rarely is a design really well suited for multiple objectives. It is important to be specific about survey goals because it is very difficult to design a good study to address a vague goal, such as "to assess the mussel community." Careful thought about what your goals really are will make it easier to choose a good design and can prevent disappointment when the data are finally analyzed.

(2) Definition of the target populations. At the end of the study, what place, time, and taxonomic or demographic group do you want the results to relate to? Do you want to make conclusions about a species living bank-to-bank in a river reach or would it be sufficient to draw conclusions about a portion of the river reach, such as the portion that is in wadable water? Are you interested in results that relate to a snapshot in time or a series of times? Are you making conclusions about all ages in a mussel population or restricting inference to a segment of the population, perhaps those visible at the substrate surface? Defining the target population spatially, temporally, taxonomically, and demographically is part of setting the goals of the survey; however, we set it aside as a separate consideration to emphasize its importance. To draw valid conclusions, sampling must coincide with the target populations—sampling must extend over the targeted place and time and include the targeted taxonomic or demographic groups; otherwise, the sampling plan missed the target. This seems an obvious requirement, but it is often not met in real mussel surveys. Survey planning will help ensure that sampling occurs at the appropriate scale and extent to match the targeted place, time, and taxa.

The identified target population may not be fully accessible because of logistical problems. For example, a stretch of river might be on private property and the property owner refuses to grant access, or a part of the river may be accessible only by boat and there is not enough time or equipment to allow for boat access. Limited access causes a mismatch between the target population and the population that can be sampled. Access problems occur regularly but should be avoided whenever possible because limited access prevents you from drawing conclusions about the target population. Because the target population and survey objectives are intertwined, limited access also prevents you from meeting the initial study objectives. Limited access is a practical problem that restricts the scope of inference. Unfortunately, there is no sampling design that can correct an access problem. If limited access is unavoidable, then the target population and the survey objectives must be adjusted.

Suppose a river-wide survey was planned with the objective of estimating density of freshwater mussels throughout the river but the only accessible sites were located at bridge crossings. Initially, the target population was the whole river, and the goal was to estimate density of freshwater mussels throughout the river. However, a large portion of the river was inaccessible, and it would be unwise to extrapolate observations made at bridge sites to the whole river, so the target population must be reduced from the "whole river" to "that portion of the river near bridge sites." Also, the survey objective must be reduced from estimating density throughout the river to estimating density near bridge sites.

(3) Resources of the investigator. How much time and labor can be devoted to the survey? Are special resources (e.g., scuba divers, taxonomically trained surveyors, boats) available? Survey goals must be realistically matched with available resources. If resources are insufficient to meet the goal of the survey, then additional resources must be mustered (and a carefully thought-out survey plan may be useful in justifying a request for additional resources) or the goals of the study must be scaled back.

(4) Characteristics of the study site. Does the study site or target species have special characteristics that will affect the performance of the survey? Such characteristics may affect the suitability and cost of various sampling devices and, therefore, dictate the sampling design. For instance, visual searches cannot be done in highly turbid waters, PONAR grabs are useless in cobble substrates, and scuba diving may be too costly or unsafe in deep, swift waters. Likewise, the size and habits of a mussel species will affect the suitability of visual searches for surveys and the amount of excavation required to estimate density.

(5) Prior knowledge about the mussel populations. Often an investigator will know something about the study population before a formal survey is begun. For example, you may know (or suspect) that mussels are found chiefly near the banks or that large mussel beds occur at known sites in the study area. (However, be careful about assuming habitat associations that have not been verified; mussels are often found in unexpected places and, conversely, not found where expected.) Prior information is immensely valuable in designing surveys because it can be used to set up stratified or adaptive designs, which can be much more powerful than purely random designs. In fact, it often will be useful to conduct preliminary, informal surveys to gather information before designing the formal survey.

An important question. Once a survey design is sketched out, a final question should be answered to assess whether the survey plan is adequate: "Will the results of the survey meet its goals?" (Figure 1B). In particular, will the survey results be precise enough to provide useful information toward the goals? Because most mussel populations are so patchy, data collected in mussel surveys usually have a large variance. This variance can be so large that the survey data are essentially useless in addressing the survey goals.

Two important steps can reduce the likelihood that this unhappy situation will develop. First, the survey goals should be stated in quantitative terms. For example, instead of simply saying that you are interested in detecting the temporal change in the brook floater *Alasmidonta varicosa* population in the Neversink River, New York, think hard about how large of a change you want to detect. Are you monitoring to detect subtle changes in population size (say ±10% between surveys), or will you be satisfied to know if the population has changed by a factor of three? Likewise, if your survey objective is to know what species are present in a reach of river, you should give some thought to what it means when a species is not found at a study site. Unless the study area is quite small, it can be very difficult to know with certainty that a species is absent, so you may need to define "present" as "detectable given *X* effort" or "having a population of greater than *T* adults."

Second, you should check to see if the proposed survey would meet these quantitative goals. Using power analysis (see "Assessment of Precision and Power before Implementing a Survey"), it is possible to assess the probability that your design actually will be able to detect the 10% shift in the brook floater population that interests you. Likewise, methods presented by Green and Young (1993) and related methods can be used to determine if sampling effort is likely to detect species presence with sufficient certainty. Even if you are uncomfortable in stating your goal in quantitative terms, it still is useful to examine the precision of your proposed survey.

So, the first step is to quantify your goals, and the second step is to check and see if the proposed survey would meet these quantitative goals. If the proposed survey does not deliver enough precision, statistical power, or certainty, then you need to change the survey goals, redefine the target population, or increase available resources (i.e., time, money, people) allotted to the survey before redesigning the survey (Figure 1B). In some cases, you may need to abandon the survey altogether. If you omit this final step, you run the risk of finding that you have spent your money and time on a futile endeavor.

Summary

We outline the considerations and questions that should be addressed before you even put your feet in the water to study a freshwater mussel. The first order of business is a thorough assessment of the goals of the survey. Why is the survey being done, and what do you hope to accomplish? Be specific. Next you need to define the population of interest—what statisticians call the target population. Defining the target population is not simple because it must be defined spatially, temporally, demographically, and taxonomically. To be realistic and practical in your planning, you must consider the resources of the investigator. Otherwise, there could be a mismatch between survey goals and resources available to accomplish the task. If resources are insufficient then either the goals need to be adjusted or resources need to be increased. Prior knowledge about the study site and the mussel populations will help improve the survey design. These considerations will help decide how best to design and implement the survey. Finally, you should ask an important question: Will the results of the survey meet its goals? You should ask this question in quantitative terms. Because survey design can be an iterative process, asking this important question gives you the opportunity to change your survey design, if need be, to ensure the survey yields the information you need to manage and conserve freshwater mussel populations.

Sampling Designs

It is useful to think of a survey as a sequence of four steps (Figure 1C): (1) choosing a sampling design, (2) collecting mussels within the sampled areas, (3) observing attributes of the collected mussels, and (4) analyzing observations (data) to estimate population-level attributes. Each activity is affected by the considerations discussed above—survey goals, target populations, available resources, and prior knowledge. Several options are available at each step. We discuss the four steps in order.

The first step in the process of conducting a survey is to select where mussels will be collected and observed. Ultimately, mussels will be collected from particular locations in a study area. Because it is uncommon for mussels to be collected from the entire study area, we must consider designs to select or sample specific places where mussels will be collected and observed. There are many sampling designs, and some are better suited for mussel surveys than others (Table 1). We consider sampling designs that are good choices for mussel surveys and those that may not be good choices but are commonly applied in mussel surveys. We will try to steer you away from the poor choices. A good sampling design can reduce the cost and effort of a survey while increasing the quality and applicability of the resulting data, while a poor sampling design can negate all the hard work done in the field and result in essentially useless data.

Complete Coverage of a Study Site

Occasionally, it will be possible or necessary to cover the entire study site and attempt to collect all the mussels living there. Visual searches by scuba or snorkeling are the usual sampling methods of choice for complete censuses. Because search speeds usually are on the order of 3–30 m^2/h (Strayer et al. 1997; Smith et al. 2000), complete coverage is slow and expensive. Because of this restriction, complete coverage is used only in cases where small and highly localized impacts are anticipated or for specialized research projects. For instance, complete coverage of a site on the Cacapon River, West Virginia, provided results useful for a computer-simulation-based comparison of sampling methods (M. M. Salehi, Isfahan University of Technology, and D. R. Smith, unpublished data). Even in these circumstances, though, the census is not truly complete because small or

TABLE 1. Cross-tabulation of survey designs and common objectives for sampling freshwater mussels. Survey designs are combinations of sampling designs and collection methods. We exclude survey designs that are unlikely to be applied (e.g., complete coverage with excavation). We indicate the survey designs that we recommend (R), recommend with reservation (RWR), or do not recommend (NR) under common objectives.

					Common objectives				
Sampling design	Collection method	Common name for design	Result	Cost	Species presence	Population density or size	Detecting an impact	Monitoring species presence	Monitoring population density or size
Complete coverage[a]	Superficial visual/tactile search of substrate surface[b]	Qualitative sampling or timed search	Species list	High	NR	NR	NR	NR	NR
	Detailed visual/tactile search of substrate surface[c]	Semi-quantitative census	Surface density and species list	High	NR	NR	NR	NR	NR
Informal sampling[d]	Superficial or detailed visual/tactile search of substrate surface[b]	Qualitative sampling or timed search	Species list	Low	RWR[e]	NR	NR	RWR[e]	NR
Probability-based sampling	Detailed visual/tactile search of substrate surface within sampling units[c]	Semi-quantitative sampling	Estimate of surface density and species list	Medium	R	NR	NR	R	NR
	Excavation of sampling units[f]	Quantitative sampling	Estimate of total density	High	NR	R	R	NR	R

[a] Extremely time-consuming; useful only for research purposes and very small sites.

[b] Search efficiency and proportion visible at substrate surface varies with place, time, and taxon. Variation in search efficiency and proportion at surface could confound spatial, temporal, and taxonomic comparisons.

[c] Proportion visible at the substrate surface varies with place, time, and taxon. Variation in proportion at the surface could confound spatial, temporal, and taxonomic comparisons.

[d] Unlikely to yield a representative sample. There is no valid way to measure uncertainty.

[e] Validity of qualitative sampling is determined by amount of search effort and coverage of the site. The relationship between species detection and search effort in qualitative sampling is poorly understood, but see Metcalfe-Smith et al. (2000) for an empirical study.

[f] Excavation of sediment is time-consuming and potentially damaging to mussels and their habitat. Amount of excavation needed to estimate total abundance could be reduced by use of double-sampling design.

buried animals are overlooked. Complete census by excavation and sieving (which is truly complete) is used very occasionally in research studies focused on very small areas (<10 m^2; e.g., Downing et al. 1989).

Informal Sampling

In an informal sampling design, places for sampling mussels or sediments within a site are chosen without a formal design for the convenience of the investigator. (Statisticians call this approach "haphazard sampling.") Thus, a biologist may search for mussels where the habitat looks promising, where visibility is good, where access is convenient, and so on. Such designs are used widely by field biologists because they require little or no planning, are flexible, and are easy to execute in the field. Examples include most timed searches, in which investigators use superficial visual searches to locate mussels at convenient places (e.g., riffles near bridges), or quadrats or transects deployed in convenient or interesting places. Despite their attractiveness and wide use, informal designs have a very serious flaw—it is impossible to make any inferences about an entire mussel population from informal samples without making the unproven assumption that the samples are representative of the target population. Usually, informally selected samples will *not* be representative of the larger population of interest, so data collected in an informal design will be biased to an unknown extent. Also, there is no valid method to assess sampling variance from informal sampling unless questionable assumptions are made. Consequently, results from informal samples are reported without measures of uncertainty and often are given greater credence than deserved. Many biologists seem to believe that data collected using informal designs are reliable for assessing population density, relative abundance of species across species or across sites, and temporal changes in mussel populations. Unfortunately, this belief is not supported by data; in fact, the few existing data strongly suggest that data collected using informal designs are of very limited use (e.g., Figures 17 and 19). Because of these problems, informal sampling is most useful in preliminary surveys and for determining the presence of a mussel species at a site and should be avoided for other applications (e.g., estimating population size, relative abundance, or temporal changes of mussel populations).

Probability-Based Designs (Design-Based Inference)

The remaining sampling methods that we discuss can all be described as "probability-based methods" because the methods allow for the estimation of sampling probabilities, which are then used to estimate a population parameter (e.g., density or abundance) and the variance of the estimate. Most methods determine the probability of sampling a mussel by first determining the probability of sampling a unit of sediment where the mussels might be found (also known as a sampling unit). A couple of the methods, namely mark–recapture and distance sampling, estimate the probability of sampling a mussel directly. However, these latter methods seem to perform poorly when applied to low-density populations of freshwater mussels. Because malacologists most often are interested in species that could be rare at a site, we favor probability-based methods that focus first on sampling a unit of sediment. There are many kinds of probability-based methods (see Thompson 1992 for a good overview); we present a few of the more common designs.

Simple Random Sampling

In simple random sampling, the spatial area of interest (e.g., the entire river bottom within the range of a species; the area of the lake bottom within 100 m of a proposed sewage outfall) is divided into nonoverlapping units. The number of units is often denoted as N. The investigator then randomly selects some number n of these units for collection and observation of mussels. In practice, all N possible units are numbered consecutively and then n numbers between 1 and N are drawn from a random-numbers table (such tables are given in most statistics textbooks; alternatively, most statistical software or calculators generate random numbers). Alternatively, each unit can be assigned x- and y-coordinates and two random numbers can be selected, which correspond to the x- and y-coordinates of the units to be sampled. Box 1 gives an example of a simple random design.

Simple random designs produce unbiased estimates of mussel population densities and other attributes but often are very inefficient. That is, because mussel populations are highly clumped at several spatial scales (Figure 3; Downing and Downing 1992), many of the random samples will contain no mussels and just a few will contain many mussels (Figure 4). Consequently, estimates of mussel population size or density may be imprecise (have large error bars) and not be very useful. This problem will be especially severe for rare species, which often are of most interest to investigators. There are two classes of solutions to this problem: (1) increase the area sampled by increasing the size or number of units sampled or (2) adopt a design that retains a random element but which focuses sampling effort on areas where mussel densities are highest. You may need to implement both approaches to achieve desired results.

Systematic Sampling

Systematic selection of habitat units is a useful alternative to simple random selection because it is easy to implement under field conditions, provides good spatial coverage of the area of interest, and is efficient for rare and clustered populations (Christman 2000). Also, systematic sampling was found to be a better design than simple random sampling for describing spatial distribution of mussels because systematic sampling spreads out the sampling effort and provides good spatial coverage of the study area (P. S. Pooler, Virginia Tech University, and Smith, unpublished data). In systematic sampling, units (e.g., quadrats) are selected

BOX 1. An Example of Simple Random Sampling

Suppose we want to estimate the number of mussels living in a short reach of stream, 10 m long by 5 m wide. We divide the study area into N = 50 nonoverlapping 1-m^2 squares (Figure 2). The location of each square can be described by its *x*, *y* coordinates; we decide to set (0,0) as the lower left corner of the reach and define the location of each square as the distance of its lower left-hand corner from (0,0).

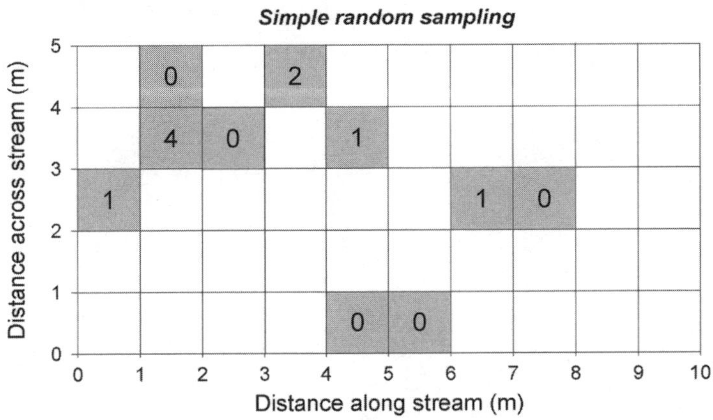

Randomly selected coordinates (x,y)

(6,2)	(4,0)
(5,0)	(3,4)
(0,2)	(7,2)
(1,3)	(2,3)
(1,4)	(4,3)

FIGURE 2. A simple random sampling design. A 10-m-long reach of stream selected for study was divided into 50 possible sampling units. Ten of these (shaded in gray) were selected using a random-numbers table and surveyed for mussels. The numbers of mussels found in each sampled unit are shown.

We decide that we want to sample n = 10 of the squares (based on the time we have available and the statistical precision needed to meet our objectives). We go to a random-numbers table and choose 10 numbers between 0 and 9 (the *x*-coordinates of the selected squares) and 10 numbers between 0 and 4 (the *y*-coordinates of the selected squares). These pairs of random numbers, which are listed to the right of Figure 2, define the locations where we will take our samples (the gray squares in Figure 2).

We go out to the stream on a fine summer day, use stakes and a tape measure to find the selected sampling sites, and search each of them for mussels, finding the numbers of mussels shown in the figure. The mean density of mussels is simply the arithmetic mean of these counts (0.9), and the estimated number of mussels in the study area is 0.9 mussels/m^2 × 50 m^2 = 45 mussels. The standard errors (SE) of these estimates are 0.41/m^2 and 20.3 mussels, respectively. To calculate the 90% confidence limits on our estimates, we note that we have $n - 1 = 9$ degrees of freedom. The value of t for $p = 0.1$ and 9 degrees of freedom is 1.833 (available in most statistical textbooks), so our 90% confidence interval is

$$0.9 \pm \text{SE} \times t = 0.9 \pm 0.41 \times 1.833 = 0.9 \pm 0.75.$$

Likewise, the confidence interval on the number of mussels is 45 ± 20.3 × 1.833 = 37.3. We conclude that the mean density and number of mussels in the study area are 0.90 ± 0.75/m^2 and 45 ± 37 mussels, respectively.

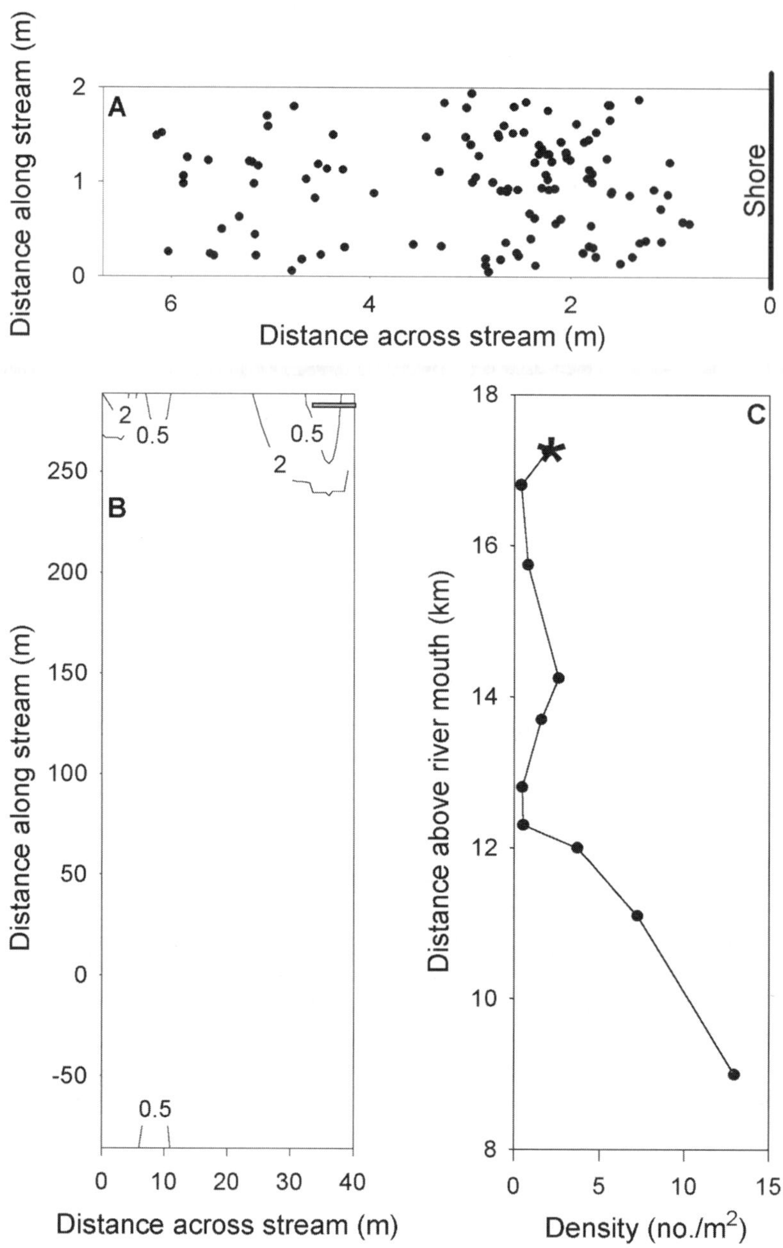

FIGURE 3. Clumping in mussel populations in the Neversink River, New York, at different spatial scales. (A) Locations of individual mussels (black dots) in a small plot. (B) Locations of mussel beds in a reach of stream. Contour lines indicate densities of mussels (number/m^2). The small gray box in the upper right shows the location of the small plot that is diagrammed in panel A. (C) Densities of mussels in 10 reaches of stream. Reaches are 100–400 m long; the reach detailed in panel B is marked with an asterisk. All data were collected using intensive visual searches (Strayer, unpublished data).

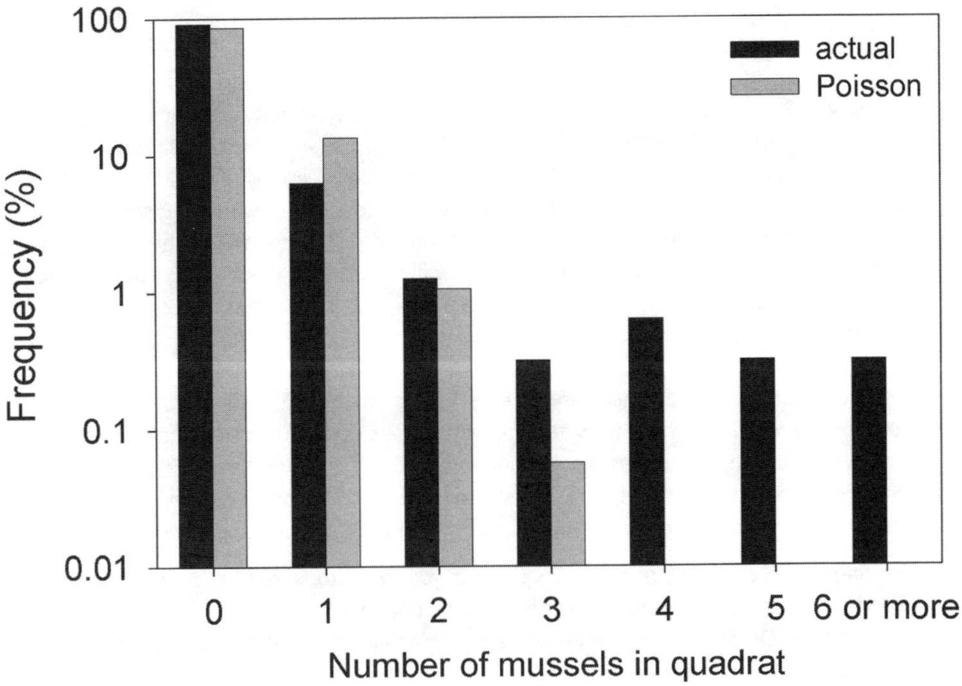

FIGURE 4. Frequency histogram of mussel densities in 319 1-m^2 quadrats placed in the study site shown in Figure 3B, compared with that expected from a Poisson distribution (expected if the mussels were randomly distributed). Note the high number of quadrats containing no mussels and the higher-than-expected number of quadrats containing several mussels. Note that the *y*-axis is log-scaled. Data from Strayer (1999b).

at regular distances (intervals) from a random starting point; each unit that follows from a random start is within (or is part of) the same systematic sample (Box 2). The distance between samples is adjusted to achieve a desired sample size and provide for good spatial coverage. If more than one random start is incorporated in the design (we recommend three random starts), then sampling variance can be estimated without making any assumptions about the spatial distribution of mussels. In the case of multiple random starts, each systematic sample is a sampling unit, and the number of multiple starts determines the sample size. (The sampling units within a systematic sample are not independently selected; if one of its units is selected, then all are selected. Because of this dependence, each systematic sample contributes only one to the sample size. Multiple random starts and, therefore, multiple systematic samples are needed for sample size to be greater than one and for variance to be calculated properly. If there is only one random start, then you must assume that mussels are randomly distributed in the study area to estimate variance.)

BOX 2. An Example of Systematic Sampling

Let's revisit the mussel population in Box 1 to introduce systematic sampling with multiple random starts. Systematic sampling is simply the process of selecting sampling units at equal intervals. Systematic sampling results in a chain or grid of sampling units that make up one "systematic sample." It is a good idea to start a systematic sample at a random location or, in other words, at a "random start." It is an even better idea to include *multiple* random starts so that there will be multiple systematic samples and variance can be estimated properly (see pages 16–17 for further discussion on variance estimation under systematic sampling).

Systematic sampling is appealing because it results in good information (i.e., precise estimates and good spatial coverage) and is easy to implement in the field. To describe the procedure, we outline the steps to systematic sampling below and then we apply the procedure in a step-by-step fashion to the example population. To make the outline specific, we refer to a quadrat as the sampling unit, but you could substitute another sampling unit, such as transect or even site, in place of quadrat depending on your sampling situation.

Step 1.—Decide the number of quadrats you want to sample and the number of random starts (three random starts is a good choice),

Step 2.—Determine the interval or distance between quadrats within a systematic sample,

Step 3.—Generate a random location for each random start and place a quadrat at each of the random starts, and

Step 4.—Sample the site by placing quadrats at equal intervals beginning with the random starting locations.

Step 1.—Suppose in our example we set out to systematically sample the mussel population from Box 1. We decide to select at least 10 quadrats, $n = 10$, and to include three random starts. (Typically, you would base the sample size on how precisely you want to estimate density. A sample size table for quadrats is provided in Smith et al. 2001:127.)

Step 2.—Next, figure out the interval between quadrats. Here is a handy formula for calculating the distance (or interval) between systematically selected units:

$$d = \sqrt{\frac{L \cdot W}{n/k}},$$

where d is the distance between units, L and W are the length and width of the study site, n is the total number of quadrats, and k is the number of random starts. The unit of measurement for d will be the same as for L and W. In our example population, $L = 10$ m, $W = 5$ m, and $n/k = 10/3 = 3.33$. Thus, $d = 3.87$ m. We then round *down* to $d = 3$ m to ensure that at least 10 quadrats will be sampled. (Note that, in this example, we rounded the distance down to the nearest meter because the quadrat is 1 m on each side. However, if we were using a 0.25-m^2 quadrat, then we would round the distance down to the nearest half-meter because that is the length of each side of a 0.25-m^2 quadrat.) The formula calculates distance for both directions in a two-dimensional survey. So, the

(Box continues on the following page.)

BOX 2 (continued)

quadrats within each systematic sample will be 3 m apart in both the across-river and upriver directions.

Step 3.—Next, we need to generate random coordinates for the three random starting locations. When sampling along one dimension (e.g., positioning cross-river transects along a riverbank), then we would need only one random number to locate a random start. However, in our example, we are sampling in two dimensions (e.g., placing quadrats a certain distance up and across the river), thus, we need two random numbers to locate a random start. For convenience, we will start sampling at the lower-left corner of the site; however, you could start in any corner of the site. Because the interval is 3 m, the random starts must be located in the 3-m by 3-m area in the lower-left corner of the site. To specify each random start, generate two random numbers between 0 and 3, then round down to the nearest integer. In this case, each random number will be 0, 1, or 2. (If we were using 0.25-m^2 quadrats, then we would round the random numbers down to the nearest half-meter.)

Suppose the first two random numbers turn out to be 0 and 2. Then the coordinates for the first random start would be (0, 2) or, in other words, the first of the three starting quadrats will be located 0 m from the left edge of the site and 2 m up from the bottom edge of the site. We would repeat this process to find the second and third random starts. Suppose the random coordinates turn out to be (0, 2), (1, 0), and (2, 2), then the three starting quadrats will be located at these three random locations.

Step 4.—The other quadrats in the systematic samples will be located at 3-m intervals from the starting locations. As a result, 13 quadrats will be sampled. In Figure 5, we show the quadrat locations.

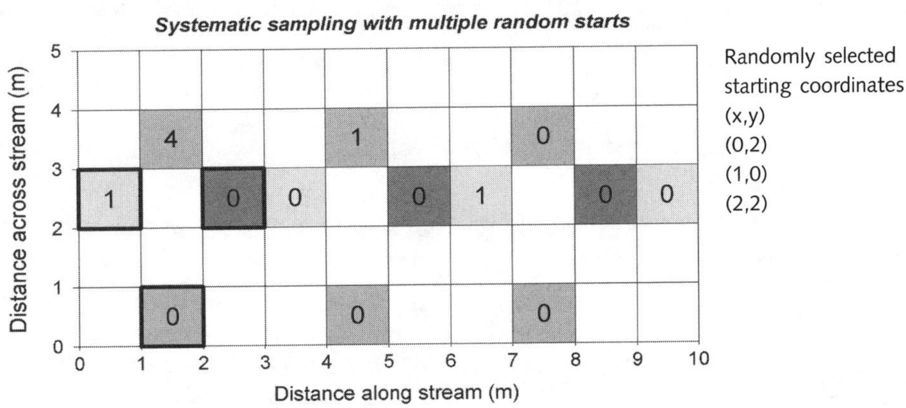

FIGURE 5. A systematic sampling design. A 10-m-long reach of stream selected for study was divided into 50 possible sampling units. Three random coordinates (marked by heavy borders) were generated to serve as starting locations. Additional quadrats were selected at 3-m intervals. All quadrats that originate from a random starting coordinate are part of one systematic sample. The three systematic samples in this example are different shades of gray. The numbers of mussels found in each sampled unit are shown inside that unit.

(Box continues on the following page.)

BOX 2 (continued)

There are three systematic samples (m) because of the three random starts. The number of possible systematic samples (M) is nine within the 3-m by 3-m area where the starting coordinates were selected. The estimate of population size ($\hat{T}$) is the average count per systematic sample ($\bar{x}$) multiplied by the number of possible systematic samples. The quadrat counts are shown in Figure 5. Thus,

$$\hat{T} = M\left(\frac{\sum_{i=1}^{m} x_i}{m}\right) = 9\left(\frac{2+5+0}{3}\right) = 21.$$

To estimate population density (number/m^2), simply divide the population total by the area (A) of the site,

$$\hat{\mu} = \frac{\hat{T}}{A} = \frac{21}{50} = 0.42/\text{m}^2.$$

Estimates of variance are based on random sampling of the systematic samples. An estimate of variance for population size is

$$\begin{aligned}\hat{\text{var}}(\hat{T}) &= \frac{M(M-m)}{m}\cdot\frac{\sum_{i=1}^{m}(x_i-\bar{x})^2}{m-1}\\ &= \frac{9(9-3)}{3}\cdot\frac{[(2-2.33)^2+(5-2.33)^2+(0-2.33)^2]}{3-1}\\ &= 114.\end{aligned}$$

An estimate of variance for population density is simply variance for population size divided by area squared. That is,

$$\hat{\text{var}}(\hat{\mu}) = \frac{\hat{\text{var}}(\hat{T})}{A^2} = \frac{114}{2500} = 0.0456.$$

Based on simulations, we have found that estimates of mussel population size or density tend to be skewed right in distribution. As a result, the usual approach to confidence interval calculation may be inaccurate. We recommend two alternative approaches

(Box continues on the following page.)

BOX 2 (continued)

to calculate confidence intervals. One approach to confidence interval calculation is based on a logarithmic transformation of the estimate and a delta-method approximation of variance (Seber 1982),

$$\exp\left(\ln(\hat{T}) \pm t_{\alpha/2,\,\mathrm{df}}\sqrt{\frac{\hat{\mathrm{var}}(\hat{T})}{\hat{T}^2}}\right).$$

Using this approach with the example data, a 90% confidence interval for population size is (5, 93). The other approach is a based on a modified bootstrap procedure presented by Hedayat and Sinha (1991:360–362). A standard bootstrap procedure is not appropriate for finite population sampling designs (such as those presented in this monograph) because units are sampled without replacement. Hedayat and Sinha modified the bootstrap procedure for sampling without replacement, which made it applicable for the sampling designs that we present. (Several other authors have recommended modified bootstrap procedures for finite population sampling situations. Christman and Pontius [2000] evaluated several modified bootstrap procedures for use with adaptive cluster sampling.) Using the modified bootstrap procedure with the example data, the 5th and 95th percentiles of the bootstrap distribution of the estimate of population size are 6 and 36. The percentiles of the bootstrap distribution are analogous to a confidence interval; thus, a 90% bootstrap confidence interval is (6, 36). Clearly, the two approaches lead to very different upper confidence limits in this case. Procedures for estimating confidence intervals for mussel populations is an area that should receive more study. However, we expect the two approaches outlined above (log-based and bootstrap-based intervals) are reliable when sample size is at a more realistic level—when quadrats number in the hundreds rather than the tens.

Systematic sampling can be thought of as a type of cluster sampling (Thompson 1992) where each systematic sample is analogous to a cluster. Cluster sampling, in general, is efficient when cluster means are similar but units within clusters vary widely. This condition tends to be met for spatially patchy populations because systematic sampling distributes sampling units widely through the study area. Thus, some quadrats within a systematic sample will hit mussel beds and others will not, but these highs and lows average out so that the multiple systematic samples are similar, and sampling variance is thus reduced.

Systematic sampling can be applied at multiple spatial scales, to select sites in a large-scale survey or to select units within a site, for example. To select sites for a broad-scale survey by systematic sampling, first number the sites falling within the area of interest, then generate a random number between 0 and the interval

(k) between systematically selected sites. The random number identifies the first selected site. Additional sites are selected at equal intervals, that is, select every kth site after the random start. The initial random site plus all sites selected at regular intervals constitute one systematic sample. When applying systematic sampling, you should include multiple random starts to estimate variance and guard against the interval between samples corresponding to a periodicity in the mussel density (e.g., riffle-pool spacing).

To systematically select sampling units (e.g., quadrats) within a site, the site is set up as for simple random sampling; that is, the site is divided into N units, which are either numbered consecutively or given x, y coordinates (Box 2). An initial sampling unit is chosen with a random-numbers table, just as for simple random sampling. The next sampling unit is found by simply adding the chosen value of the interval k to the initial random number; additional sampling units are chosen using this procedure until you reach the end of the study area. If the sampling units are given x, y coordinates, as often is convenient, the upriver and across-river intervals do not have to be the same. However, Pooler and Smith (unpublished data) found that setting these intervals to be equal is a robust approach, and setting the upriver interval smaller than the across-river interval should be avoided because of parallel-to-the-current density patterns that seem typical for mussels. The unit at the initial random coordinate plus all units that follow at regular intervals constitute one systematic sample. Multiple random starts are required for multiple systematic samples and for variance estimation. Box 2 shows an example of data and calculations for a systematic survey design.

Because units are located at regular intervals (after a random start), systematic sampling is easy to implement in the field. Locations of consecutive units within the same systematic sample are on bearing (either an upriver or across-river bearing) and separated by the length of the interval. Depending on water depth, flow, and interval between units, we have found different methods useful for locating the positions of sampling units. These have included 1-m and 2-m rods, ropes cut to the interval length and long (50–100 m) lines with marks (e.g., surveyor's tape) to indicate quadrat locations. Units are kept on bearing by use of a compass or similar device. Long lines deployed along the upriver bearing are efficient because the line can be moved quickly to locate quadrats throughout the site.

Software to estimate species-specific density and abundance based on systematic sampling with multiple random starts can be found at *http://www.lsc.usgs.gov/aeb/2068/index.asp*. The software also includes algorithms and information useful in survey design, such as sample size tables, computation of intervals between units for a desired sample size, and generation of random coordinates.

Stratified Sampling

If you know (or suspect) where mussel populations are densest, you can concentrate your sampling effort on those areas by using a stratified sampling design.

First, you divide the study area into two or more strata based on known or suspected mussel densities. Note that strata can be defined by criteria other than mussel density. For example, if scuba diving is much more costly than snorkeling, you could divide the sampling area into low-cost (shallow water) and high-cost (deep water) areas. Stratified sampling is flexible because each stratum can be sampled using a different sampling design and sampling effort. Any of the designs that we discuss can be applied independently within each stratum. For example, Smith et al. (2003) applied adaptive cluster sampling (ACS) in a low-density stratum and conventional sampling in a high-density stratum.

The strength of a stratified design arises because the investigator can devote different amounts of sampling effort to each stratum, based on the mussel population expected and the cost of sampling in each stratum. Stratified sampling is efficient when the number of mussels per sample is similar across samples taken *within* any single stratum but very different for samples taken *across* strata (i.e., strata means differ greatly). For mussel surveys, this condition tends to be met when the area of interest can be stratified by levels of density and cost of sampling.

If the cost of sampling is the same in all strata and simple random sampling is applied in all strata, the optimum allocation of sampling effort (i.e., the one that gives estimates of density or abundance with smallest variance) is

$$n_h = \frac{nN_h\sigma_h}{\sum_{k=1}^{L} N_k\sigma_k},$$

where n is a fixed total sample size, N_h is the number of units in the hth stratum, σ_h is the population standard deviation for the hth stratum, L is the number of strata, and n_h is the sample size for the hth stratum (Thompson 1992).

If sampling costs differ across strata and total cost (c) is fixed, then the optimum allocation is (Thompson 1992)

$$n_h = \frac{(c - c_0)N_h\sigma_h/\sqrt{c_h}}{\sum_{k=1}^{L} N_k\sigma_k\sqrt{c_k}},$$

where c_0 is a cost that is common to all strata regardless of sample size (call it setup cost) and c_h is a stratum-specific cost per sampling unit. Estimates of variances and costs in each stratum could come from past samples, from preliminary samples taken specifically to set up a stratified design, or from the investigator's general knowledge and intuition about the study area. Note that population variance can be predicted from population density and the size of the sampling unit using the following empirical relationship (Downing and Downing 1992):

$$s^2 = 1.49m^{1.168},$$

where m is the mean number of mussels per sampling unit (i.e., the density in number/m^2 times the area of the sampling unit in m^2). Thus, if sampling cost does not vary across strata, the best allocation of samples is approximately

$$n_h = \frac{n N_h m_h^{0.584}}{\sum_{k=1}^{L} N_k m_k^{0.584}},$$

and if sampling costs vary across samples, the optimal allocation of samples is approximately

$$n_h = \frac{(c - c_0) N_h m_h^{0.584} / \sqrt{c_h}}{\sum_{k=1}^{L} N_k m_k^{0.584} \sqrt{c_k}}.$$

These last two equations may be useful for planning in cases where the investigator does not have any direct estimates of population variance but can guess at the expected population density of mussels in different strata. Box 3 shows a stratified survey design.

Stratified random designs can give much more precise estimates of mussel populations than simple random designs, given the same sampling effort. Because stratified random designs can be so much more efficient than simple random designs, it often is worthwhile to do some preliminary sampling to define strata and allocate sampling effort. Christman (2000) concluded that stratified random sampling is the best design to sample rare and clustered populations when using quadrats and that systematic sampling is a good second choice if preliminary information is not available to define strata boundaries.

Stratification can also be used to delineate areas of special interest or to facilitate comparisons, regardless of efficiency considerations. This application of stratified sampling is especially useful in large-scale surveys. For example, in a river-wide survey, the river might pass through regions of different land use or geology. An investigator who wishes to compare mussel populations among regions could use regions to form strata and allocate the number of sites among strata to assure precise stratum-specific estimates.

We offer a word of caution about applying stratified random sampling to a long-term monitoring study. Stratification is efficient for mussel surveys when the study area is partitioned into strata with similar density. Over the long term, the spatial distribution and densities of mussels could change, making the initial stratification scheme inefficient. Instead of basing the stratification scheme on a variable that may change during the life of the program, it may be better to use a more robust sampling design (e.g., systematic sampling) or to be prepared to change the stratification scheme as conditions change.

BOX 3. An Example of Stratified Sampling

Let's return to our hypothetical 10-m by 5-m reach of stream for which we want to estimate mussel density or numbers. In this case, suppose that we know from preliminary samples (or even an informal walk through the study reach) that most of the mussels are clustered along the left ascending bank and that we can guess that mussel densities along this bank are ~10 times higher than those elsewhere in the reach. We, therefore, define two sampling strata: one containing all of the units adjoining the lower bank and the second containing all other units (Figure 6). Again, we decide to sample a total of n = 10 units. Using optimal allocation theory,

$$n_h = \frac{nN_h m_h^{0.584}}{\sum_{k=1}^{L} N_k m_k^{0.584}},$$

where n is the total number of units to be sampled; n_1 and n_2 are the number of units to be sampled in the high- and low-density strata, respectively; N_1 and N_2 are the total number of units in each of the L = 2 strata; and m_1 and m_2 are the mean densities in the two strata. Substituting the numbers from our example, we find that

$$n_1 = \frac{(10)(10)(10^{0.584})}{[(10)(10^{0.584}) + (40)(1^{0.584})]} = 4.9,$$

$$n_2 = \frac{(10)(40)(1^{0.584})}{[(10)(10^{0.584}) + (40)(1^{0.584})]} = 5.1.$$

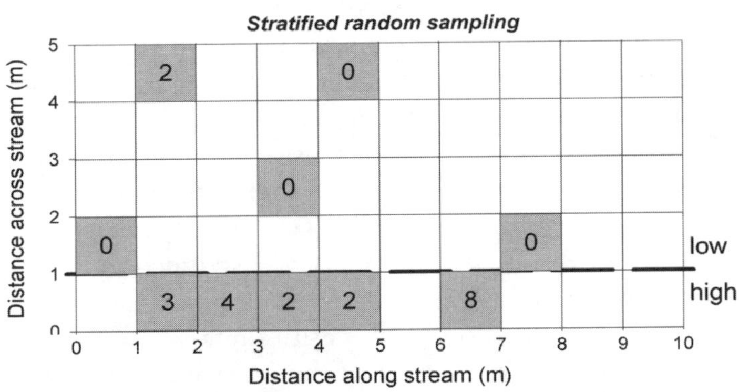

Random coordinates

High	Low
3	(1,3)
1	(0,0)
6	(4,3)
4	(3,1)
2	(7,0)

FIGURE 6. A stratified-random sampling design. A 10-m-long reach of stream selected for study was divided into 50 possible sampling units. Two strata were defined: a high-density stratum with 10 units and a low-density stratum with 40 units. Five units from each stratum were selected using a random-numbers table (shaded in gray) and surveyed for mussels. The numbers of mussels found in each sampled unit are shown.

(Box continues on the following page.)

BOX 3 (continued)

Therefore, we decide to randomly sample five units within each stratum and come up with the numbers of mussels shown in the figure. The mean density in the study area is the mean in each stratum weighted by the size of the stratum:

$$\bar{x} = (0.2)(3.8) + (0.8)(0.4) = 1.08/\text{m}^2.$$

We estimate the total number of mussels in the study area to be $1.08/\text{m}^2 \times 50\ \text{m}^2 =$ 54 mussels. Many statistics texts give formulas for estimating variance and confidence intervals for stratified designs; we note that estimating the number of degrees of freedom is not straightforward and was addressed by Cochran (1977) and Krebs (1998).

Double Sampling

In double sampling, the investigator measures a response variable in a large number of sampling units using an approximate, inexpensive method, and resurveys a fraction of the units with a precise but costly method. The design relates the results from the inexpensive method to those from the costly method to achieve precise estimates at less cost than if the costly method were applied alone. Double sampling is flexible and can be applied at multiple spatial scales. For example, it can be used to select either sites or units within sites.

The double-sampling and stratified-sampling designs can be combined in what is termed "double sampling for stratification" (Cochran 1977; Thompson 1992). Double sampling for stratification is useful when you believe that stratification might be helpful but you do not know the boundaries or mussel densities of the strata. For example, in large-scale mussel surveys such as a river- or drainage-wide survey, it may be useful to stratify river reaches by density; however, information may not be available beforehand to stratify sites. An inexpensive, approximate survey can be used to gather the information needed to stratify the reaches for a more costly and precise survey (Box 4). As is the case for stratified random designs, the fraction of plots selected for intensive survey will ordinarily differ among strata, with much more effort expanded on high-density strata.

Another use for double sampling is in calibrating an approximate method with a more precise method (Eberhardt and Simmons 1987; Smith et al. 2000). Probably the most obvious example for mussels would be calibrating counts of mussels that are visible at the sediment surface against the total number of mussels (visible plus buried) actually present in the sampling unit (Box 5). In this case, double sampling is applied to the smallest sampling unit (e.g., the quadrat). The approximate method is applied on a large representative sample of units, and the precise method is applied to a representative subset of the initial sample. An easy and effective way

BOX 4. Double Sampling for Stratification to Assess Mussel Populations Over a Drainage Basin

Because mussel biologists have not yet published examples of this method, we present a fictional example. Suppose you want to estimate the numbers of a rare mussel species in a large watershed. Using your knowledge about the species' biology, you judge that the species is unlikely to be found in streams smaller than fourth order. The study area is, therefore, defined as all streams in the basin larger than fourth order. You then divide all streams in the study area into roughly equal-area reaches (say, 1,000 reaches of 2,000 m^2 each), give each an identification number, and randomly select 200 of the reaches for study. You conduct 1-h timed searches on each of these reaches, with the following results.

Mussel populations	Number of sites
No mussels of any kind present	80
Mussels present, but target species not seen	80
Target species present at catch rates of <5 animals/h	35
Target species present at catch rates of >5 animals/h	5

You have time to do quantitative surveys at only 20 sites, so you decide to concentrate effort where the mussels are. Therefore, you decide to sample all 5 of the putative high-density sites, 8 of the low-density sites, 4 of the sites with mussels but no target species, and 3 of the sites with no apparent mussels. The sites you sample within each density-class are randomly chosen. At each of the 20 sites, you use your favorite sampling design (let us say 210 0.25-m^2 quadrats placed in a systematic design with three random starts), with the following results.

Site	Density-class	Observed density of target species (number/m^2)
1	No mussels	0
2	No mussels	0
3	No mussels	0
4	Mussels, but no target species	0
5	Mussels, but no target species	0
6	Mussels, but no target species	0.08
7	Mussels, but no target species	0
8	Low density	0
9	Low density	0
10	Low density	0.33
11	Low density	0
12	Low density	0
13	Low density	0
14	Low density	0.08
15	Low density	0
16	High density	0.83
17	High density	0.08
18	High density	0.42
19	High density	0.42
20	High density	0.17

(Box continues on the following page.)

BOX 4 (continued)

From Thompson (1992), we calculate the basinwide mean density as

$$\bar{y}_d = \sum_{h=1}^{L} w_h \bar{y}_h ,$$

where $\bar{y}_d$ is the estimate of the basinwide mean density, w_h is the proportion of sites in each of the *h* strata, and $\bar{y}_h$ is the mean density within each stratum. In our case,

$$\bar{y}_d = (0.4 \times 0) + (0.4 \times 0.02) + (0.175 \times 0.05) + (0.025 \times 0.38) = 0.026 \text{ mussels/m}^2.$$

The estimated number of mussels in the basin is then just this density multiplied by the area in streams larger than fourth order, or $0.026/\text{m}^2 \times 2{,}000{,}000 \text{ m}^2 = 52{,}000$ mussels.

to assure representativeness is to select the subset systematically; for example, select every third unit in the sample and implement the precise method in addition to the approximate method. The relationship between approximate and precise measurements can be modeled with a ratio estimator or regression. If the relationship between the approximate and precise estimators is thought to be a line passing through the origin, then a sample ratio can be calculated:

$$r = \frac{\sum_{i=1}^{n} y_i}{\sum_{i=1}^{n} x_i} ,$$

where y_i is the total number of animals collected using the precise method (e.g., visual searches plus sieved sediment) and x_i is the number of animals collected using the approximate method (e.g., visual searches by themselves). The density or number of animals in the target population is simply calculated as the number or density estimated using the approximate method times this ratio. Thompson (1992) gave further details on calculations, including formulas for variance.

If the relationship between precise and approximate methods is thought not to pass through the origin, a regression estimator is more appropriate than a ratio estimator. Here, the counts from the precise method are plotted against those from the approximate method, and a linear regression is fitted through the line. This regression is then used to calibrate the counts from the approximate method.

BOX 5. Using Double Sampling to Include Buried Mussels in a Population Estimate

Many mussels live buried out of sight in the sediments and so escape ordinary visual searches. It usually is too time-consuming (and perhaps too damaging to the habitat) to search through all the sediments in the study area for these buried mussels. Double-sampling offers a practical way to gather sound information on the entire mussel population (including buried mussels) that is not too time-consuming.

Our example is taken from Smith et al. (2001), who were interested in estimating the size of mussel populations in and around the site of a bridge replacement in the Allegheny River, Pennsylvania. Because Smith et al. were interested in the entire mussel population and because the site was suspected to contain clubshell *Pleurobema clava*, a legally protected species that is known to burrow deeply, Smith et al. needed to include mussels buried out of sight in the sediments. The study area was defined to be a 100-m-long by 188-m-wide (the width of the river) reach of river. First, Smith et al. used systematic sampling with multiple random starts to select 562 0.25-m^2 quadrats for visually searching (they chose this sample size based on the expected density of the target species and the desired precision of the results). Then, following the guidelines shown in Table 2, they decided to excavate a third of the quadrats to a depth of 10 cm and sieve them through a 6.35-mm mesh, which should have captured all of the adults that were buried in the sediment. To assure that the excavated quadrats were representative of the larger sample, quadrats were selected systematically for excavation, that is, every third quadrat was excavated. They found 26 clubshells at the sediment surface in total. In the 183 excavated quadrats, they found 4 visible and 9 buried clubshells. Using a ratio estimator, we estimate that the ratio of the true number of mussels in a quadrat to those that are visible is

$$r = \frac{\sum_{i=1}^{n} y_i}{\sum_{i=1}^{n} x_i} = \frac{13}{4} = 3.25 \ .$$

Thus, we conclude that the actual number of clubshells present in the 562 quadrats was $3.25 \times 26 = 84.5$, which corresponds to a density of 0.60/m^2. Smith et al. (2001:131) used a slightly more complicated regression estimator to conclude that the density of this species was 0.38/m^2, with a standard error of 0.087.

Smith et al. (2000, 2001) gave further details on use of regression methods in double sampling, including formulas.

The optimal allocation of effort to the initial approximate survey and the subsequent precise survey depends on the costs and precision of the two methods (Thompson 1992). For a fixed total cost, the lowest variance in the estimate of population density or size is obtained when

$$\frac{n}{n'} = \sqrt{\frac{c'}{c}\left(\frac{s_r^2}{s^2 - s_r^2}\right)},$$

where c and c' are the costs of sampling a unit by the precise and approximate methods, respectively, n and n' are the number of units sampled by these two methods, s^2 is the variance in the precise samples, and s_r^2 is the variance in the relationship between the precise and approximate estimates of the population

$$s_r^2 = \frac{1}{N-1}\sum_{i=1}^{N}(y_i - rx_i)^2,$$

where N is the number of sampling units in the study area, x_i and y_i are the approximate and precise estimates of the population in sampling unit i, and r is the sample ratio:

$$r = \frac{\sum_{i=1}^{n} y_i}{\sum_{i=1}^{n} x_i}$$

(that is, the sum of precise estimates divided by the sum of approximate estimates in plots where both were estimated). Smith et al. (2000) collected information on time-costs of visual searches and excavation (using a 6.35-mm mesh sieve) and concluded that the optimal allocation of effort depended on the percentage of the target population that was buried out of sight (Table 2).

Double sampling is especially useful where the target variable (e.g., a mussel population) is highly patchy, where the details of this patchiness are not known prior to the survey, and where there are alternative ways to assess the population that differ greatly in cost. As these conditions often are met in mussel surveys, double sampling has great potential in surveys of mussel populations. It probably

TABLE 2. Suggested rules for determining the optimal proportion of quadrats to excavate in a double-sampling design, from Smith et al. (2000).

% of population visible at surface	Optimal proportion of quadrats to excavate
>60%	0.25
50–60%	0.33
40–50%	0.5
<40%	1.0

will be especially useful in dealing with the vexing problem of estimating the numbers of mussels that are buried out of sight beneath the sediment surface. So far, it has not been used widely (but see Smith et al. 2001 for a good example).

Adaptive Sampling

Another way to concentrate sampling effort where mussel populations are densest is to use an adaptive sampling design. In an adaptive design, sampling units are initially laid out using a conventional sampling design (i.e., simple random, stratified random, systematic). If the response variable (e.g., mussel density) in a sampling unit exceeds some predetermined threshold, then the investigator takes additional samples in the vicinity (or "neighborhood") of this sampling unit. Adaptive sampling is easy to do in the field and is a good way to focus effort on mussel beds in cases where the precise location of the beds is unknown prior to sampling. Even in cases where large-scale patchiness is known and is accommodated through use of a stratified design, the details of smaller-scale patchiness often are unknown and so can be dealt with using an adaptive design. Initial results using adaptive designs to sample mussel populations have been disappointing (Strayer et al. 1997; Smith et al. 2003). Nevertheless, adaptive designs probably deserve to be considered for mussel surveys and ecology.

Adaptive designs were described in detail by Thompson (1992) and Thompson and Seber (1996), from which the following account is summarized. For simplicity, we present an example of adaptive sampling based on a simple random design, but it can be readily combined with other sampling designs, such as stratified or systematic designs. First, units to be sampled are selected using a simple random approach. Next, the investigator chooses the criterion that will trigger additional sampling around these initially selected units. This criterion indicates that the response variable in the sampling unit is interesting enough to the investigator to merit further sampling in the area. Usually, the criterion is the presence of a rare species or a high mussel population density, but it is extremely flexible and could include such things as the presence of appropriate habitat (e.g., gravel substrate, overlying canopy) or an indicator species that the investigator thinks is associated with the target response variable. Once the criterion for further sampling is set, it must remain fixed for that study site, so it pays to think carefully before setting a criterion. It often is useful to spend a few minutes cruising the sampling site before settling on a criterion. (The criterion for further sampling may differ among sampling sites. Thus, you may set the criterion at 1 mussel/quadrat at a site where mussels are scarce and 12 mussels/quadrat at a site where mussels are abundant.)

Once the initial sampling units and criterion for further sampling are chosen, sampling commences. If the response variable in a sampling unit fails to meet the criterion, the investigator moves on. If the response variable meets the criterion, the investigator then samples all adjacent sampling units (actually, the choice of the secondary sampling units to be sampled is more flexible than this [Thompson 1992; Thompson and Seber 1996] but normally, adjacent units are sampled). If any of these sampling units meets the criterion, units adjacent to them are sampled and so on until no additional units are found to meet the criterion (Box 6). Then, the investigator moves to the next primary sampling unit. When the sampling is complete, the investigator has been able to concentrate his or her effort on the areas where mussels exist, instead of barren areas and, thus, theoretically, can produce more precise estimates of population density than nonadaptive designs, given the same effort. In addition, adaptive designs allow the investigator to encounter and take biological observations (e.g., size, reproductive status) on more individuals of the target species than would a simple random design. The special formulas needed to estimate population density or size and their variances for adaptive designs were given by Thompson (1992) and Thompson and Seber (1996).

Adaptive sampling was designed for populations of rare, clustered organisms and so seems well suited to mussel populations. Nevertheless, neither Strayer et al. (1997) nor Smith et al. (2003) could detect any increase in efficiency of adaptive designs when applied to real mussel populations in the field. Thus, the method should be regarded as experimental at this point. Further, we have encountered three difficulties in using adaptive sampling for mussel populations. First, it is unclear what value to set for the criterion for further sampling, in a theoretical sense. Should the top 5% of the primary sampling units trigger further sampling, or would it be better to aim for 15% or 30%? We have aimed for ~10%, but there is no theoretical justification for this number. Second, if the criterion is set too liberally, it is possible to get into a cluster that is very large (even as large as the whole sampling area), and which requires excessive sampling effort. At present, there is no way for an investigator to break off sampling in a cluster of units that has become too large to be practical to finish. Adaptive cluster sampling is an area of active research, and techniques to prevent excessive sampling effort are receiving particular attention (see Smith et al. 2003 and citations therein). Finally, calculations for estimating population variance in adaptive designs are fairly complex. Software to compute estimates based on ACS with an initial random sample can be downloaded from *http://www.lsc.usgs.gov/aeb/davids/acs/index.asp.*

Two-Stage Sampling

Often the spatial extent of the target population is so large that habitat must be sampled in two stages—first, primary units (e.g., large habitat units such as river reaches) are selected and then secondary sampling units are selected with the primary units (e.g., quadrats within river reaches). This class of sampling method is called two-stage sampling. Mostly, we have discussed methods to sample units within a site. However, double sampling for stratification, which we discussed above, is an example of a two-stage design.

BOX 6. An Example of Adaptive Sampling

We return to our example of trying to estimate the number of mussels in a 10-m by 5-m reach of stream (Figure 7). In the simplest form of adaptive sampling, we select random coordinates as starting points. Using the same methods as in simple random sampling (Box 1), we select three units (again, a smaller number than we would use in the real world, to keep the example simple) for sampling (Figure 7). Before we begin formal sampling, we spend a few minutes to examine the study reach to get an idea of the mussel populations we are likely to encounter. We then set the condition to trigger additional sampling at two mussels/m^2; that is, if we find two or more mussels in any of the randomly chosen units, we sample the four units immediately adjacent.

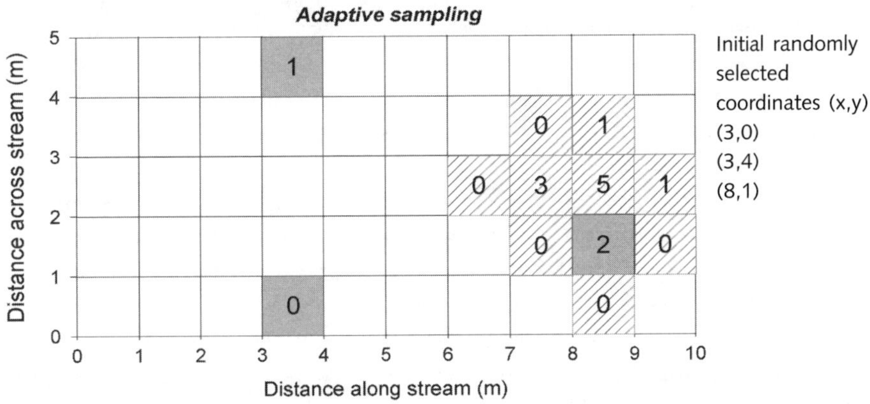

FIGURE 7. An adaptive sampling design. Primary sampling units are shown in gray; additional units sampled as a result of the adaptive design are crosshatched. The number of mussels found in each unit is shown inside that unit. See Box 6 for explanation.

When we sample, we find that only one of the three primary units (specifically, the unit at 8, 1) meets the condition of two or more mussels. We sample all the units adjacent to this unit and find that one of *these* units has five mussels and, therefore, meets the condition for further sampling. We then sample the units adjacent to the unit with five mussels and find that one of *these* units meets the condition. We continue to sample until none of the units at the margin of our sampled cluster meets the condition of two or more mussels. The results of all this sampling are shown in Figure 7. Using the formulas for estimating the mean (Horvitz-Thompson estimator) and variance for adaptive cluster sampling given by Thompson (1992), we estimate that the study reach contains 1.49 mussels/m^2, or 74.6 mussels in all. The standard errors are 1.95/m^2 and 97 mussels, respectively.

Two-stage sampling is flexible, allowing the application of any probability-based method at either the first or second stage. In practice, we caution against making the design overly complex. For the first stage of sampling, we recommend application of a straightforward method, such as simple random sampling, stratified random (including double sampling for stratification), or systematic sampling. A wider variety of methods can be applied at the second stage of sampling. You could apply whichever probability-based method you deem appropriate and efficient to sample within a site.

Large-scale population estimation would be a typical objective of a survey that uses two-stage sampling (e.g., estimates of river-wide or basin-wide population size or density). Variation in the estimate has two components, one for each stage of sampling. Precision of the large-scale population estimate is determined mainly by how many sites are sampled rather than how many quadrats are sampled within a site. If the percentage of sites selected for sampling is less than 10%, then the contribution to overall variance from within-site sampling is negligible (Salehi and Seber 1997; Seber 2001). This suggests that sampling effort should be concentrated on sampling as many sites as possible rather than on taking many samples at each site. However, sampling costs must be considered and the cost of sampling a site is usually considerably more than the cost of sampling within a site. Nevertheless, for large-scale (e.g., river-wide) surveys, it is important to sample as many sites as you can afford.

Recently, adaptive sampling has been combined with two-stage sampling in hopes of increasing the efficiency of two-stage sampling (Salehi and Seber 1997; Salehi and Smith, unpublished data). This design, called two-stage sequential sampling, shows promise for freshwater mussel surveys where two-stage sampling is justified by cost, for example, in river-wide surveys or site surveys in large rivers (Salehi and Smith, unpublished data). To apply two-stage sequential sampling, large primary units are selected at random. Within each primary unit, a sample of secondary units is selected at random. If a criterion is met within a primary unit, then additional secondary units are selected. Two-stage sampling is an attractive option for large sites, especially where scuba diving is required. Advantages of two-stage sampling are mostly logistic; for example, positioning a dive boat is costly, so that limiting the numbers of times that the dive boat must be relocated could reduce the survey cost. In Box 7, we present an example of two-stage sequential sampling to a freshwater mussel population.

Probability-Based Designs (Model-Based Inference)

Probability-based sampling designs can be classified as "design-based" or "model-based." Most of the sampling designs that we have considered (i.e., simple random, stratified, systematic, adaptive, double sampling, and two-stage) are "design-based" because random selection of units as specified by the sampling design is all that is required for accurate estimation of population parameters such as density and abundance. No underlying distributions or models were assumed. (Double sampling for calibration is an exception because it can include a regression model in the estimation, but the modeling requirements are modest.) Thus, design-based approaches lead to nonparametric population estimates because accuracy of

BOX 7. Using Two-Stage Sequential Sampling

Two-stage sampling is an attractive option for large sites, especially where scuba diving is required. In two-stage sequential sampling, which combines two-stage sampling with a type of adaptive sampling, the site is subdivided into areas of equal size, and areas are selected at random. Within each selected primary unit, a predetermined number of quadrats are selected at random. If the species is found within a selected area, then additional quadrats are selected randomly. In this way, more sampling effort is allocated to where the species is found in the site.

A 40-m reach on the Cacapon River was completely covered, and a detailed search of the surface was conducted. Three species were present: eastern elliptio *Elliptio complanata*, northern lance *E. fisheriana*, and yellow lampmussel *Lampsilis cariosa*. Eastern elliptio was the most numerous and relatively clustered; northern lance was rare and clustered; and yellow lampmussel was rare but not particularly clustered. These populations were used to evaluate two-stage sequential sampling through computer simulation (Salehi and Smith, unpublished data).

The site was divided into 48 areas, each 5 m by 5 m. At the first stage, several 5-m by 5-m areas are selected, and, at the second stage, 0.25-m^2 quadrats within each 5-m by 5-m area are selected. If any mussels are found within a 5-m by 5-m area, then additional quadrats are selected at random. Salehi and Smith (unpublished data) found that two-stage sequential sampling was more efficient than conventional two-stage sampling for eastern elliptio as long as at least six quadrats were sampled initially in each 5-m by 5-m area. Gains in efficiency were greater for northern lance with variance from sequential sampling typically about 90% of variance from conventional sampling. There was not much gain or loss in variance for yellow lampmussel.

Two-stage sequential sampling is a promising approach for sampling sites on large rivers. We plan to evaluate the method further by applying it to a site on the Allegheny River.

estimates depends on random selection of sampling units, not on underlying statistical models or the spatial distribution of mussels within a site.

Another class of sampling design does rely on an underlying model or models, and these are called, not surprisingly, "model-based" designs. Dorazio (1999) reviewed the theory and application of model-based designs to mussel surveys. We introduce a regression model-based design but add to the discussion mark–recapture and distance sampling, which are model-based in that they make use of probability models to estimate populations.

Regression Designs

In theory, at least, it may be possible to predict mussel density in a sampling unit from some more readily measured variable (e.g., water depth). In such a case, an investigator could sample both mussels and water depth in a small number of sampling units, measure the water depth in a large number of sampling units (even a complete census), then estimate the number of mussels in these sampling units. This approach is most attractive when the response variable (e.g., mussel density) can be precisely predicted by a variable or variables that can be obtained at much lower cost. At present, regressions between easily measured variables and mussel density have been very imprecise, so regression approaches have not been used widely in mussel ecology. Nevertheless, they may be useful in some circumstances, especially for variables other than mussel density. Thompson (1992) should be consulted for more information on regression designs.

Mark–Recapture Models

Mark–recapture models are used widely in ecology (for review articles see Pollock et al. 1990; Skalski and Robson 1992; Schwarz and Seber 1999; Williams et al. 2001) but have rarely been used to assess mussel populations (however, see Hart et al. 2001; Rogers et al. 2001). In mark–recapture designs, investigators catch a number of animals from the target population, mark them, and release them back into their habitat. Later, the target population is resampled (perhaps multiple times), and the numbers of marked and unmarked animals are recorded. Either the whole study site is searched for mussels to achieve a representative sample or the study site is partitioned into units and units are randomly selected for searching. Skalski and Robson (1992) discussed use of mark–recapture methods in combination with survey sampling designs.

Mark–recapture models are classified into two categories by whether recapture occurs while population size is static (termed "closed-population models") or after the population has been changed by birth, death, immigration, and emigration (termed "open-population models"). In a closed population, you can estimate the total number of animals (i.e., population size) by knowing the number of marked animals and the ratio of marked to unmarked animals (Chao 2001). If animals are marked and recaptured multiple times when populations are open to recruitment, mortality, and migration, then population size and rates of population change (e.g., survival rate and recruitment rate) can be estimated (Pollock et al. 1990; Schwarz and Seber 1999). Closed- and open-population models have been combined to form a hybrid model, which is called the robust design (Kendall et al. 1995). In the robust design, mark–recapture sampling is done over a short period of time so that population size can be estimated using closed-population models and then sampling is repeated annually so that vital rates can be estimated using open-population models. For example, suppose that a river site is searched for mussels each day over a period of 3 d in May 2000. On each day, the site is completely searched, unmarked mussels are marked, and previously marked mussels are recaptured. It is reasonable to assume that the population is closed over the 3-d period, so that closed-population models can be used to estimate population size. Now suppose that the 3-d mark–recapture sampling is repeated in May

2001 and 2002. Not only can population size be estimated for May 2000, 2001, and 2002, but also the multiyear mark–recapture sampling can be used to estimate annual rates of survival and recruitment.

There are a number of important assumptions for proper application of mark–recapture models. Some of the assumptions depend on the design of the study—the tagging, capture, and recapture procedures. For example, tags must not be lost or become illegible, recaptures must be reported accurately, tagging must not change the survival or capture of animals, and tagged animals must be representative of the target population. Other assumptions depend on behavioral characteristics of the animals. For example, the fate of one tagged animal must not be linked to the fate of others and emigration must be permanent (i.e., animals that leave do not return during the course of the mark–recapture study). Although some of these assumptions can be tested, violations of the assumptions cannot be remedied by data analysis and will cause inaccurate population estimates. The seriousness of the inaccuracy depends on which assumption is violated and the extent of the violation (see Burnham et al. 1987 and Pollock et al. 1990 for discussions of testing assumptions and consequences of assumption violation).

Additional assumptions are made to account for variation in the rates of capture, survival, recruitment, and movement. Contrary to the assumptions discussed in the previous paragraph, these assumptions can be incorporated into the mark–recapture model and form the basis for a wide variety of models. For example, capture rates can be assumed to be constant through time or within an identifiable class of animal, such as species, size, and sex. This assumption leads to a model that includes a constant capture rate, and that model can be used to estimate both the capture rate and population size. The process of estimating capture rate from a mark–recapture model is analogous to the commonly used technique of least-squares regression. In least-squares regression, the estimates for the regression parameters are found by minimizing the summed differences between the observations and model predictions (these differences are commonly referred to as residuals). Similarly, a parameter in a mark–recapture model (e.g., capture rate) is estimated by finding the rate that minimizes the difference between the observed recaptures and recaptures as predicted by the model. The specific mathematical techniques differ between regression and mark–recapture estimation (a technique called maximum likelihood estimation is used in the mark–recapture case), but the basic idea of finding the estimates that best fit the observations is the same.

Other models can be specified to account for variation in demographic rates. For example, capture rates can be assumed to vary by time, and a model can be specified that allows for time-specific capture rates. A large number of models can be specified, each derived from different assumptions about variation in capture rates (Otis et al. 1978; Pollock et al. 1990). There are models to account for behavioral effects of capture on capture rates and other models to account for variation in capture rates among individual animals. Each model gives different estimates of population size; thus, model selection is an important part of the analysis of mark–recapture data. Assessment of how closely the model "fits" the data, as measured by the difference between model predictions and the observed data, is used to test hypotheses about population rates and select the best models (Otis et al. 1978; Pollock et al. 1990; Burnham and Anderson 2002). Estimates of population parameters (e.g., population size) are based on the best-fitting model or models. Chao (2001) reviewed eight closed-population models (originally presented in Otis et al. 1978), each imposing a different structure on capture probabilities in terms of time, behavioral response to capture (behavior), and individual variation (heterogeneity). An example of mark–recapture sampling of freshwater mussels to estimate population size is presented in Box 8.

Analysis of open-population models can be even more complex than closed-population models because probabilities of survival, movement, recruitment, and capture are all incorporated into the mark–recapture model. Schwarz and Seber (1999) projected that the general framework for open population can lead to models with "over 1,000 parameters." However, with the availability of comprehensive, user-friendly software to analyze mark–recapture data (e.g., the program MARK [White and Burnham 1999] for open-population studies and the program CAPTURE [Otis et al. 1978] for closed-population studies), the application of mark–recapture methods is limited more by data than by methodology. Model selection has emerged as a significant problem in mark–recapture modeling because of the large number of possible models that can be fit to the data. Burnham and Anderson (2002) present modern and powerful methods of model selection; fortunately most of these methods have been implemented in user-friendly software, such as the program MARK.

Mussel populations present special problems for mark–recapture methods. Mark–recapture methods were developed largely for populations of mobile animals (e.g., fish and vertebrate wildlife) for which marked and unmarked animals can mix thoroughly between sampling occasions. Mussels are relatively immobile, so it is unlikely that marked and unmarked animals mix thoroughly between sampling occasions, especially sampling occasions that are separated by only a short period of time (i.e., a few days or weeks) as is necessary for proper application of mark–recapture methods. For example, buried mussels would be unavailable for capture if only the sediment surface is searched. However, buried mussels could return to the surface and be available for capture by the next sampling occasion. Likewise, a mussel at the substrate surface could bury itself, thus becoming unavailable for capture at the next sampling occasion. This is referred to as temporary emigration, where mussels can, in effect, leave the target population but later return. Two types of temporary emigration have been studied: completely random and Markovian (Kendall et al. 1997; Williams et al. 2001). In completely random temporary emigration, the probability that a mussel is buried is unrelated to whether it was buried previously. In Markovian temporary emigration, whether a mussel is buried depends on its vertical position at the previous sampling occasion.

BOX 8. Using Mark–Recapture Sampling to Estimate Population Size

On 20 June 1994, a three-pass removal survey was conducted at a 40-m-long pool/run of the Cacapon River at Capon Bridge, West Virginia (Smith, R. F. Villella, and D. P. Lemarié, U.S. Geological Survey, Leetown Science Center, unpublished data). The three-pass removal technique is a mark–recapture method where a defined area is searched completely three times. Each search constitutes a pass and the number of passes can be increased as time allows; the more passes, the more precise the population estimate will be. During a pass, mussels are found and marked. By marking a mussel, it is in effect removed from subsequent sampling.

To apply the method at the Cacapon River site, an aluminum nail flagged with surveyor's tape was inserted in the substrate adjacent to each mussel. A different color tape was used to distinguish between mussels found during the first and second passes. During the third and final pass, newly found mussels were counted but not marked. To enhance searching, the site was partitioned into eight transects, each 5 m wide and oriented bank-to-bank across the river. A biologist tried to search each transect for 30–40 min per pass. Searching was visual because of high water clarity, and the substrate was not disturbed during the search.

The survey took 21.9 biologist-hours to complete and resulted in 416, 214, and 152 mussels found during the first, second, and third passes, respectively. Species present were (in order of abundance) eastern elliptio, northern lance, and yellow lampmussel. The program CAPTURE (Otis et al. 1978) was used to analyze the data and estimate population size. Model M_{bh}, which assumes behavioral and heterogeneous variation in recapture rates, was used to estimate population size because it is appropriate for removal sampling (Otis et al. 1978). The resulting population estimate was 980 mussels with a standard error of 38.3 and a 95% confidence interval of 918–1,069.

In July 1994, a census (i.e., complete coverage and detailed search of substrate surface) was conducted at the site where mussels on the substrate surface were counted within all 0.25-m^2 cells bank-to-bank along the 40-m-long site. It was determined that 1,208 mussels were at the substrate surface. The mark–recapture estimate was 228 below the actual population and the confidence interval failed to include the true population total, with the upper confidence limit falling below the total by 139. So, the mark–recapture method in this case yielded a precise (coefficient of variation = 4%), but inaccurate estimate. We suspect that the large underestimate was due to imperfect detection of mussels at the substrate surface during the superficial search, which caused capture probability to be very low or zero for some mussels.

Temporary emigration violates a basic assumption of mark–recapture methods and can result in biased estimates of population parameters (i.e., population size, capture and survival rates).

We can think of two approaches to deal with temporary emigration in freshwater mussel studies: (1) apply advanced mark–recapture designs and models that allow for temporary emigration or (2) sample in such a way to prevent temporary emigration. Kendall et al. (1997; Kendall and Bjorkland 2001) applied the robust design to estimate temporary emigration. The robust design, as described above, combines sampling over a brief period of time (as in closed-population sampling) with sampling over longer periods of time (as in open-population sampling). In our opinion, the better way to deal with temporary emigration when studying freshwater mussels is to prevent it by effectively sampling both buried and visible mussels. Some amount of excavation to capture buried mussels should be done in addition to searching the sediment surface. The position of a captured mussel (on the surface or buried) should be recorded. Because risk of mortality could depend on whether a mussel is buried or at the surface (from predation, for example), it would be ecologically informative to incorporate vertical position in the capture–recapture modeling. Multistate models have been developed that allow for a transition between locations, for example, surface or buried (Brownie et al. 1993; Williams et al. 2001).

Although we know of no rigorous comparisons between mark–recapture and design-based estimates of mussel population size or density, we strongly suspect that design-based (e.g., quadrat surveys) are superior for that purpose. However, we predict that estimates of vital rates (survival, recruitment, and migration) will be increasingly important as the study of freshwater mussel ecology advances (see Hart et al. 2001 for an example). Open-population mark–recapture models are perhaps the best methods available to estimate rates of survival, recruitment, and migration. Thus, mark–recapture studies and open-population models may be important tools in future studies of mussel population dynamics. In either case, inequality in capture rates, where some mussels have a low or zero chance of capture, will lead to biased estimates of population parameters. Whether studying a closed or an open population, special care will have to be taken to ensure that all mussels in the target population can be captured, marked, and recaptured.

Distance Sampling

We introduce distance sampling by example. Suppose that an investigator searches for mussels in a stream along a line oriented perpendicular to the channel. The investigator maintains a position at or just downstream from the line and searches visually or tactilely upstream of the line. When a mussel is found, the investigator measures the perpendicular distance from the line to the mussel. Distance data are then used to estimate population density and abundance. Because the investigator maintains a position at the line, mussels closer to the line are more likely to be found, and frequency of detection drops off with distance from the line. Many models have been developed to approximate the frequency distribution of the distances and, in turn, estimate the probability of detecting mussels at different distances from the line. Given the number and length of the lines, number of encounters, and estimated sampling probability, mussel density can be estimated.

Buckland et al. (1993, 2001) discussed distance-sampling methods and presented comprehensive methods of analysis. Fortunately, a comprehensive software package was developed by Buckland et al. (1993, 2001) to support the analysis, which can be complex.

The advantage of distance sampling is in its ease of implementation in the field. Lines can start from random points along a riverbank, then an investigator can survey the line, which is oriented along a constant bearing. A rope or rod can be used to define the line on the substrate surface and guide the search. In this way, a number of parallel lines can be surveyed within the area of interest. The disadvantages are the critical assumption that all mussels on or near (say within 5 cm) the line are found, the somewhat subjective nature of modeling distance data, and the inability to model distance data when only a few mussels are found.

Comparative sampling indicates that quadrat-based sampling is superior to distance sampling for mussel surveys (Figure 8). At 14 streams in nine states, we

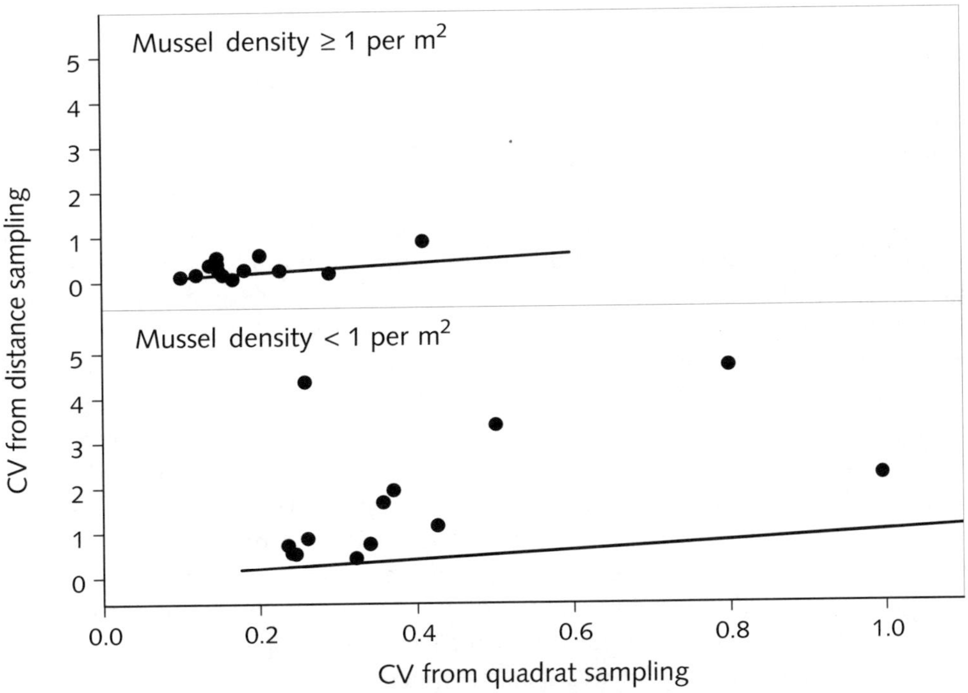

FIGURE 8. A comparison of distance sampling and quadrat-based sampling using the coefficient of variation (CV) for site- and species-specific estimates of mussels. An equivalent area was sampled using both methods at 14 sites (Smith and others, unpublished data). Panels separate sites where density was at least $1/m^2$ from those where density was less than $1/m^2$. Lines show 1-to-1 isopleths, so that circles above the line indicate that distance sampling is less precise than quadrat-based sampling, and circles below the line indicate that distance sampling is more precise than quadrat-based sampling.

applied distance sampling and quadrat-based sampling side by side to estimate mussel density (see Smith et al. 2000 for more information about the sites and species found). The coefficient of variation (CV) from distance sampling was larger than the CV for quadrat sampling for 92% (24) of the estimates (Figure 8). Thus, for a given effort, estimates of mussel density were more precise from quadrat-based sampling than distance sampling. That result, coupled with imperfect detection at the line and concern that distance sampling cannot be applied to rare species, leads us to conclude that distance sampling is not generally applicable to mussel surveys and should only be considered where densities are high ($>1/m^2$).

Choice of a Sampling Unit

When sampling within a site, the size and shape of the sampling unit should be chosen so that variation within units is high compared with variation among units, and so that the units are practical to use (i.e., fast to set up and search in the field), with the ultimate goal of minimizing variance of population estimates. The optimal size of the sampling unit depends on the spatial distribution of the target population and on costs to set up and process a sample. If the population is clumped (as mussel populations often are) and setup costs are small, then small sampling units are most efficient (i.e., give the most precise estimates of population size for a given level of sampling effort; Elliott 1977; Downing 1979, 1984). There is a limit to this recommendation because, at some point, the sampling unit will become so small that errors caused by deciding if an animal is inside the unit overwhelm any efficiency gains, but sampling units commonly used by malacologists (e.g., the 0.25-m^2 quadrat) usually are large enough to avoid this problem.

The value of using small sampling units is demonstrated by the following example. Within a 30-m × 40-m pool/run in the Cacapon River, West Virginia, all mussels that were at the substrate surface were counted and mapped to the nearest 0.25 m^2 (Figure 9). Eastern elliptio, northern lance, and yellow lampmussel were present. The *Elliptio* spp. were distributed similarly, although at different densities. Yellow lampmussel was sparser and more regularly distributed. We calculated the coefficient of variation from simple random sampling that would result from sampling the same total area with each size of sampling unit. For *Elliptio* spp., CV was directly related to sampling unit size—the larger the unit, the larger the CV (Figure 9). The advantage of a smaller unit disappeared when sampling yellow lampmussel because it was sparsely dispersed.

If the costs of setting up a sampling unit are high compared to the costs of processing the sampling unit (searching or sorting sediments), then the general recommendation to use small sampling units does not apply. Setup costs include the time spent traveling to a sampling unit, finding a sampling location, laying out the boundaries of the sampling unit, and getting people and equipment ready to take measurements at each sampling unit (e.g., suiting up divers, getting them

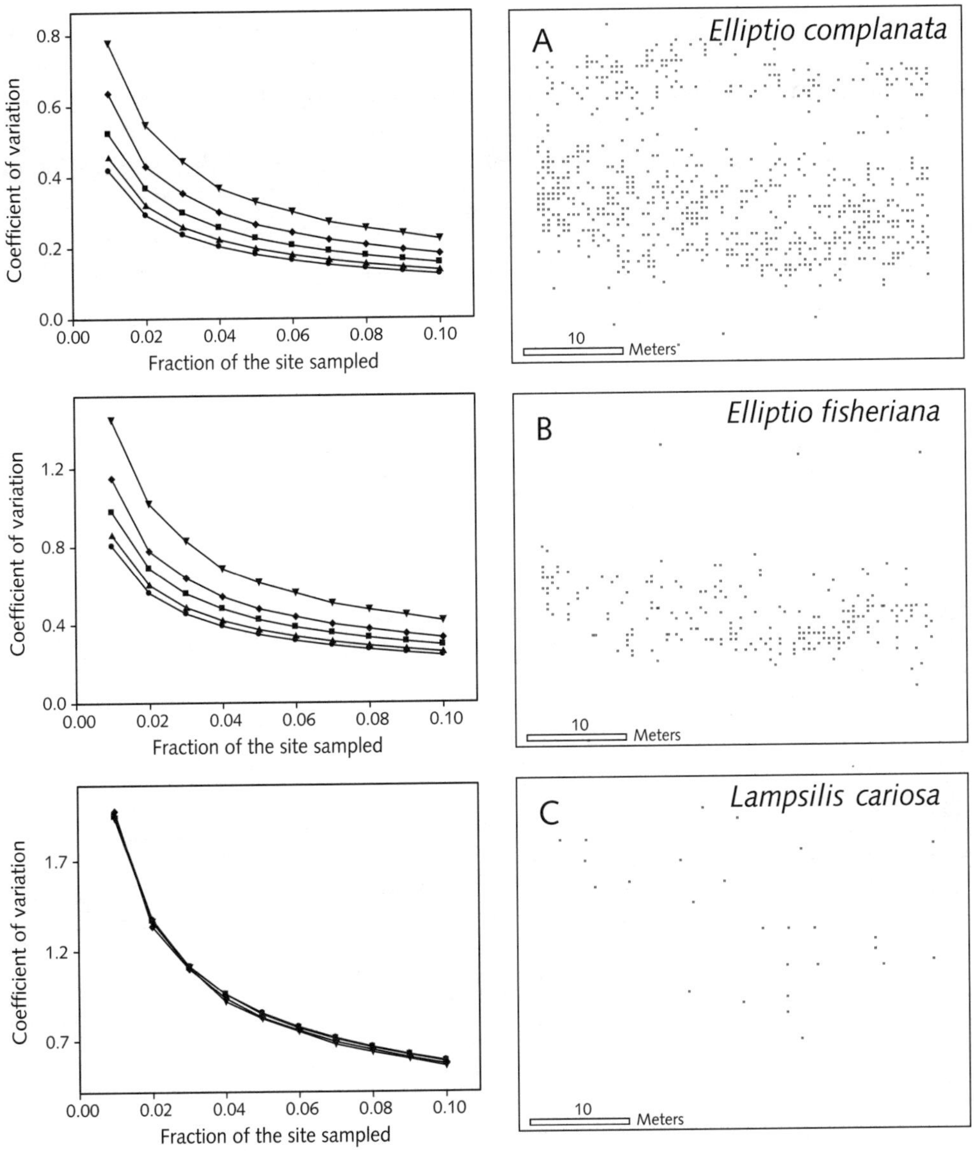

FIGURE 9. Coefficient of variation for density estimates using different size sampling units and simple random sampling. Sampling unit sizes are 0.25 m^2 (circle), 0.50 m^2 (point-up triangle), 1.00 m^2 (square), 2.00 m^2 (diamond), and 4.00 m^2 (point-down triangle). The three populations sampled were (A) eastern elliptio *Elliptio complanata*, (B) northern lance *Elliptio fisheriana*, and (C) yellow lampmussel *Lampsilis cariosa*, which were censused in a roughly 30-m × 40-m pool/run in the Cacapon River, West Virginia. Spatial distributions of mussels (shown as black dots) are shown in maps to the right of the corresponding graphs. The flow of the Cacapon River is from right to left.

into position on the sediments). Thus, particularly in large or deep bodies of water, setup costs may be substantial, and it may make sense to use large sampling units. Specifically, the most efficient size for a sampling unit is one in which the time required to search the unit is equal to the time required to set up the unit. Thus, if setup costs are high, large sampling units are more efficient than small ones (Figure 10).

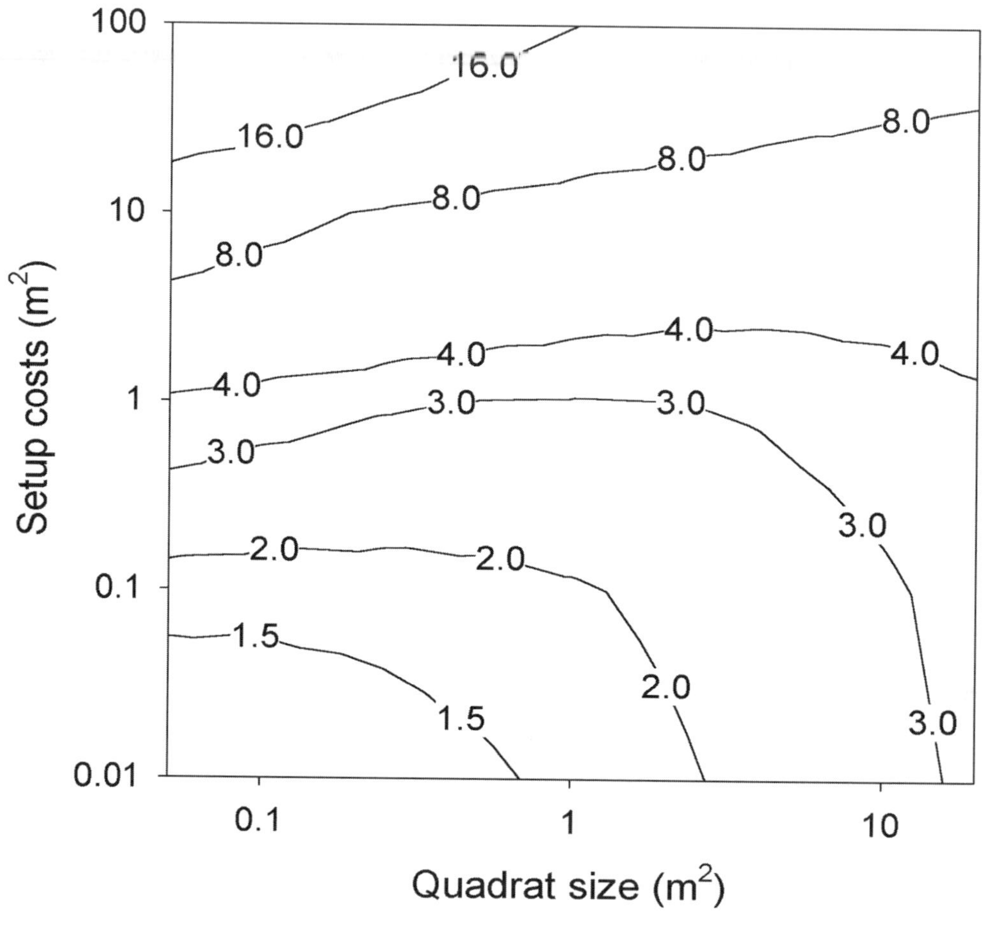

FIGURE 10. Coefficient of variation (CV) of estimated mussel density as a function of size of the sampling unit and setup costs. Setup costs are given as the area that could be searched in the time required to set up the sample. Thus, a setup cost of 1 m^2 means that it takes as long to set up the sampling unit as to search 1 m^2 of sediment. Coefficient of variation expressed in relative units, with the CV of a 0.05-m^2 sample having setup costs of 0.01 m^2 arbitrarily set to 1. Thus, if setup costs are minimal (0.01 m^2), increasing the size of the sampling unit from 0.05 m^2 to 10 m^2 nearly triples the CV. In contrast, if setup costs are high (e.g., 10 m^2), then increasing the size of the sampling unit actually decreases the CV.

Quadrats of 0.1–1 m^2 (most often 0.25 m^2) are suitable sampling units in many circumstances. They can be constructed of metal or cement-filled polyvinyl chloride pipe and are light and easy to carry. It is helpful to attach strings to the quadrat frame to define subareas within the quadrat for ease in searching (Figure 11). Quadrats are conducive to efficient sampling designs (e.g., systematic sampling) and excavation, some amount of which is necessary for rigorous density estimation (Smith et al. 2000, 2001).

Strip transects are useful in cases where larger sampling units are needed. These typically are ~1 m wide and at least 10 m long. Transects are easy to set up and are a fast way to search large areas quickly and quantitatively. They typically are set up using weighted or ordinary line fixed at both ends by anchors, stakes, or trees. It may be convenient to mark the line at regular intervals and attach a 1-m-long bar to the line to define the edges of the sampling area. Strip transects should always be oriented across an environmental gradient and never parallel to it. For instance, in small, steep-sided lakes, strip transects may be laid out perpendicular to shore to incorporate a wide variation in water depth and sediment type, which presumably are major environmental gradients that affect the mussel community (e.g., Cvancara 1972; Strayer et al. 1981; Hanson et al. 1988a). In streams, strip transects should run across the current (i.e., from one bank to the other) for ease in setup and to cross major environmental gradients. Pooler and Smith (unpublished data) found that efficiency was much lower (CV was much higher) when strip transects were oriented parallel to the channel than when oriented perpendicular to the channel (Figure 12). An important shortcoming of strip transects is that there is no practical way to excavate them, and some amount of excavation is necessary for rigorous density estimation (as we discuss below). However, strip transects oriented across the current can be useful for implementing a presence–absence survey in large rivers to assure thorough coverage of a site.

Summary

There are many sampling designs to choose from, but we recommend use of statistical or probability-based sampling designs because these designs allow rigorous inference about freshwater mussel populations. We cover the basic sampling designs and introduce some advanced designs. We assess the designs in terms of their potential application to freshwater mussel populations.

- Complete coverage of a site is considered occasionally when site area is very small or for research purposes.
- Informal sampling is a nonstatistical process for selecting where to collect and observe mussels, based on experience or convenience. Because it is nonstatistical there is no way to assess the reliability of the resulting data. We feel this procedure is applied more often than it should be in mussel surveys.

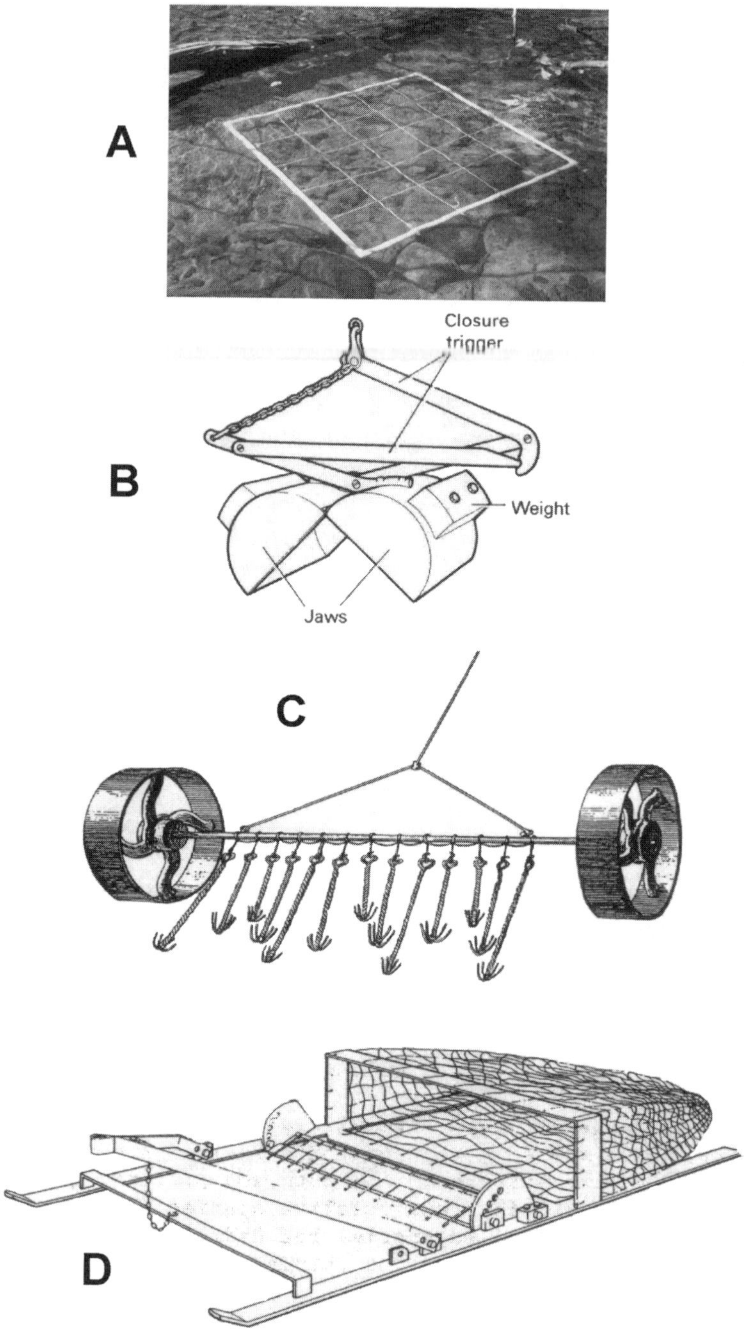

FIGURE 11. Gear useful in sampling freshwater mussels. (A) 1-m^2 polyvinyl chloride quadrat with lines at 20-cm intervals to aid in systematic searching of the quadrat; (B) PONAR grab (from Downing and Rigler 1984); (C) Brail bar, modified by the addition of wheels for use in lakes (Ortmann 1911); and (D) Benthic dredge (Miller et al. 1989).

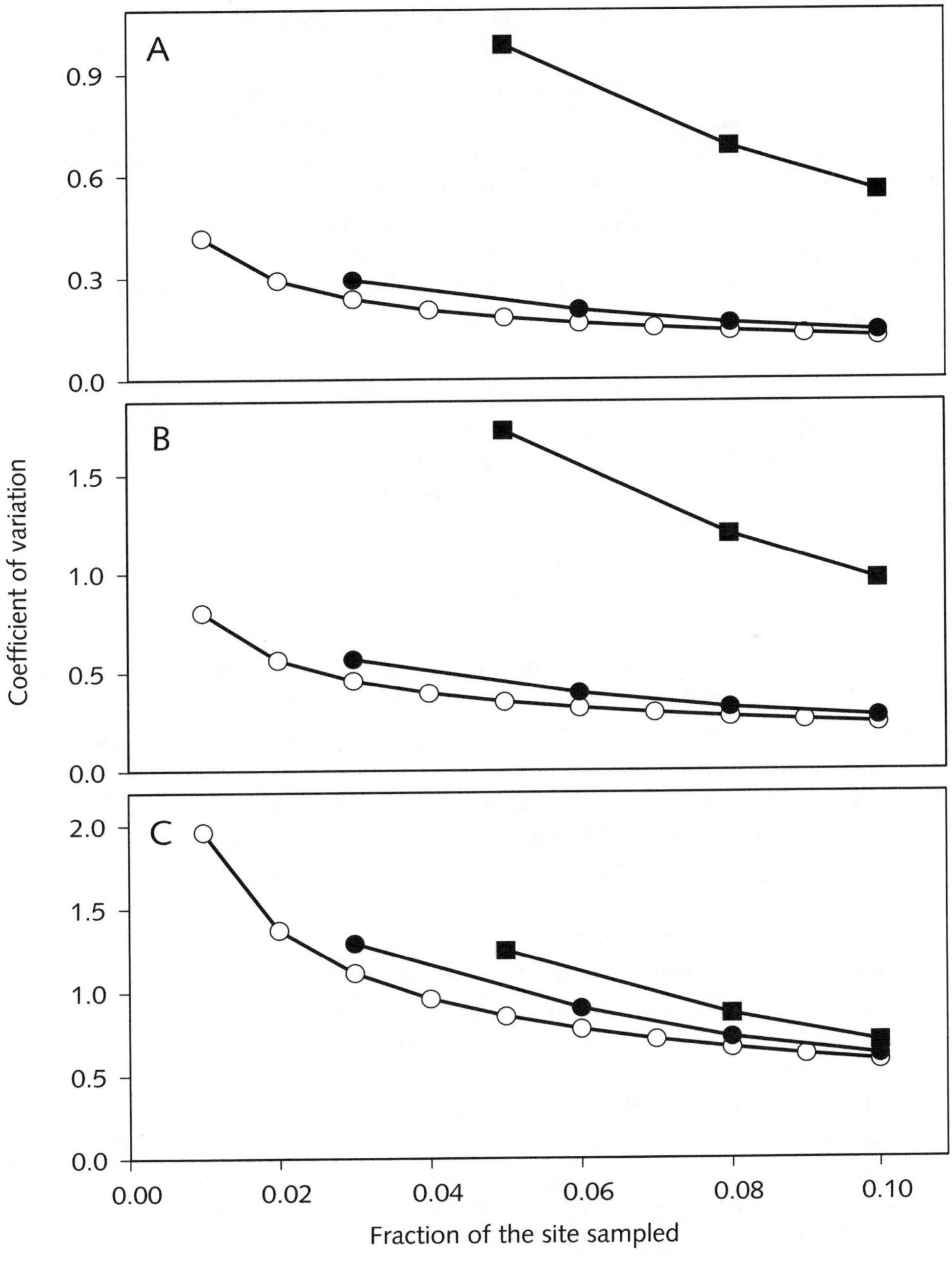

FIGURE 12. Coefficient of variation resulting from simple random sampling using a 0.25-m^2 quadrat (open circle), a 1-m-wide strip transect oriented perpendicular to the current (closed circle), and a 1-m-wide strip transect oriented parallel to the current (closed square). Populations sampled were (A) eastern elliptio, (B) northern lance, and (C) yellow lampmussel; spatial distributions of these populations are shown in Figure 9.

- Simple random sampling provides a good basis for comparison to gauge the efficiency of other sampling designs, but we do not recommend simple random sampling as a practical design for most mussel surveys.
- Systematic sampling is easy to implement in the field, it gives precise estimates for patchily distributed populations, and it spreads out sampling effort throughout the study site. Incorporating multiple random starts allows for proper variance estimation. We recommend systematic sampling for many applications.
- Stratified sampling is useful when spatial distribution of mussels is known prior to sampling. By concentrating sampling where mussels are dense, precision of estimates of density or population size can be increased.
- Double sampling is implemented in two phases with the first phase being the application of an inexpensive, approximate method of measurement and the second phase being the application of an expensive, exact method of measurement. We think double sampling holds much promise for mussel surveys, especially to include buried mussels in population estimates and designing broad-scale surveys across many sites.
- Adaptive sampling is a flexible sampling design that helps allocate sampling effort to areas where mussels are present or at high density. Although promising, we view adaptive sampling as experimental and do not recommend routine application of adaptive sampling at this time.
- Two-stage sampling is often applied out of practical necessity. For example, when conducting a survey on a large river, it is expensive to move around the site. In such a case, it may be useful to select large areas to work in, then subsample the large area using quadrats or transects. Two-stage sampling is not usually justified in terms of increased precision of estimates but rather by the high cost to move among primary units. Sampling as many primary units as possible tends to increase precision.
- Regression designs can be used to take advantage of auxiliary information, such as habitat characteristics, in selecting the sample. It should only be used when the auxiliary information is strongly correlated with mussel density, a condition that is rarely met at present.
- Mark–recapture methods can be applied to freshwater mussels. Closed-population methods assume that the population is constant during the time of the study (i.e., not subject to mortality, migration, or recruitment) and are used to estimate population size. Open-population methods allow for population change and are used to estimate vital rates (i.e., survival, recruitment, migration). Mark–recapture studies can be complex in their design and analysis, and underlying assumptions are important to their proper implementation. We do not recommend the use of closed-population mark–recapture methods to estimate population size. However, open-population methods provide a powerful set of tools for the study of vital rates, which is important for understanding the ecology and effectively conserving freshwater mussel populations.

- Distance sampling is implemented by traversing a randomly located line and recording the perpendicular distances between the traversed line and any detected mussels. The frequency distribution of the distances is modeled to estimate detection rates and, in turn, population density. Field trials of distance sampling suggest that quadrat-based sampling provides more precise estimates of mussel population size.
- The size and shape of a sampling unit should be chosen to minimize variance of the population estimate for fixed survey cost or effort and be practical to use in the field. Small sampling units usually minimize sampling error for fixed cost. However, when cost to set up a unit is high, then a larger unit would be favored. We recommend 0.25-m^2 quadrats for most situations. Strip transects, oriented across the current, may be useful for implementing presence–absence surveys in large rivers to assure thorough coverage of a site.

Methods for Collecting Mussels

Now that we have introduced methods for determining where to collect mussels, we can discuss how to collect mussels. The most common ways to collect mussels are visual searches and collection of sediments, which may be sieved. In special cases, methods such as tactile surveys, muskrat middens, brails, or dredges may be useful.

Visual Searches

The term "visual searches" covers a wide range of methods in which the observer collects mussels that he or she sees at the substrate surface. Methods range from superficial searches, where observers quickly collect mussels while wading or swimming, to painstaking, centimeter-by-centimeter searches of a stream bottom by an observer wearing a swimming mask. Superficial searches for mussels, whether the surveyor is wading or swimming, can cover a lot of ground quickly (500–5,000 m^2/h) and result in very high catch rates for visible mussels (sometimes hundreds to thousands of mussels/h; e.g., Huehner and Corr 1994; Strayer et al. 1997; Hoggarth et al. 2000). As a result, a superficial visual search, combined with informal coverage of the site, often is a good method for developing species lists or detecting the presence of rare species. Unfortunately, superficial searches and informal sampling do not provide a representative sample of the mussel community; small animals or those that bury deeply in the sediment are underrepresented (Hornbach and Deneka 1996; Vaughn et al. 1997; Obermeyer 1998; Smith et al. 2000, 2001). Thus, superficial searches and informal sampling are inappropriate for establishing absolute population sizes or densities, or even estimating the relative abundance of different mussel species or demographic groups. Searches done by snorkeling or diving presumably are significantly slower than searches done by wading and seem to be much more efficient at detecting small or cryptic mussels, although we do not know of any good published comparisons of these methods.

A detailed visual search usually is conducted within quadrats or transects to ensure that the entire sampling unit is carefully searched. Quadrats or transects can be placed in a study site using any of the sampling designs introduced previously. Detailed visual searches are much slower (typically 3–30 m^2/h) than superficial searches but result in very good recovery of all mussels that are visible at the sediment surface. Using a probability-based method to place quadrats (or transects) in a site and detailed visual searches of the quadrats can result in useful estimates of mussel density, at least for mussels visible at the surface at the time of the survey.

Of course, all visual searching methods are defeated by poor visibility in turbid or deep water and will not be useful if a large part of the target population is buried out of sight. The small amount of work that has been done on the effects of visibility on search efficiency suggests that moderate turbidity may have little effect on the efficiency of visual surveys (Strayer et al. 1997). Often, a large part of a mussel population is buried out of sight in the sediment (Amyot and Downing 1991; Balfour and Smock 1995; Smith et al. 2000). The size of this hidden fraction of the mussel population depends on life stage (juveniles are far more likely to bury than adults), season (mussels are most likely to be buried during fall and winter and most likely to be exposed when sperm or glochidia are being released; Rogers et al. 2001), mussel species, and habitat. Buried mussels will not be found even in detailed visual searches unless the sediment is excavated.

Sediment Collection

As an alternative to visual surveys, the investigator may collect bulk sediment and sieve it to separate mussels from sediment. Sediment may be collected by hand (trowel) or with a grab, and sieving may be done on site or at a more convenient location.

Sediment collection and sieving is extremely time-consuming (0.5–10 m^2/h; Smith et al. 2000; D. L. Strayer, unpublished data) but can collect all mussels, including juveniles that are difficult to sample quantitatively with other methods. As a result, it is an excellent candidate for pairing with a less time-consuming method (i.e., detailed visual search of substrate surface) in a double-sampling design (e.g., Smith et al. 2000, 2001).

In large rivers, some workers (Miller and Payne 1993; Miller et al. 1993; Richardson and Yokely 1996; Payne et al. 1997) have had divers scoop bulk sediment into a bucket, which is then taken to the surface and sieved aboard a boat or on the shore. Both PONAR and Ekman grabs (Figure 11) have been used to collect mussels in lakes and large rivers (e.g., Hanson et al. 1988a; Strayer et al. 1994). The performance of grabs has not been tested; it is likely that grabs underestimate mussel populations in some situations. Based on known characteristics of grabs (e.g., Elliott and Drake 1981), it seems likely that PONAR grabs are preferable to Ekman grabs and that standard (23 cm × 23 cm) PONARs are better than petite PONARs (15 cm × 15 cm). Nevertheless, when mussels are dense, they may get caught in the grab's jaws causing a misfire, which would tend to underestimate mussel density where populations are dense.

If the study is to include juvenile mussels, an appropriately fine mesh size should be used and sample sorting should be done in well-lighted conditions by trained, unhurried technicians. The mesh size of the sieve is chosen depending on the objectives of the survey and the grain size of the sediment; most investigators have used mesh sizes of 2–8 mm (Neves and Widlak 1987; Hanson et al. 1988b; Miller and Payne 1993; Strayer et al. 1994; Richardson and Yokely 1996; Smith et al. 2000). We note that most of these sieves still are too coarse to retain small juveniles; if small juveniles are part of the target population, then very-fine-meshed sieves must be used (e.g., Neves and Widlak used a 0.13-mm mesh). In cases where mussels are to be killed and kept for further analyses, it may be best to put the unsorted sieve residue in a cooler or freezer so it can be sorted carefully in the laboratory (e.g., Strayer et al. 1994). If sediments are sorted in the field so that mussels can be released unharmed, it is important to devote enough time and care to recover all of the mussels in the sample.

Special Methods

When visibility is too poor for visual searches, mussels can be found by touch (tactile searches), in which the investigator gently runs his or her fingers over the sediment. Although much slower than superficial visual searches, tactile searches can produce large collections of mussels if mussel populations are dense. The properties of tactile searches have not been investigated, but they presumably produce a highly biased sample of the mussel community, in which small or deeply buried animals are underrepresented, as are species that feel most like rocks. Tactile searches probably are more effective in fine-grained sediments (sand or mud) than in coarse-grained sediments (cobble). In fine-grained sediments, an inverted bow rake gently drawn over the sediment surface can be a useful tool for finding mussels (J. L. Metcalfe-Smith, National Water Research Institute, personal communication). Finally, the large amount of broken glass and other trash in many of our streams can make tactile searches hazardous to surveyors.

Muskrats *Ondatra zibethicus* sometimes eat huge numbers (>1,000) of mussels, whose empty shells they leave in neat piles (shell middens) along the stream bank. If you are trying to develop a species list for a site, muskrat middens can be an excellent source of information. They contain large numbers of shells that can be examined quickly, often include small or rare species, and are a fine source of voucher specimens. Nevertheless, muskrats do not collect a representative sample of the mussel community (with respect to species or size; e.g., Watters 1994; Tyrrell and Hornbach 1998; Smith et al. 2001), so they are of limited use for most survey purposes. Also, it is difficult to know what target population a midden collection represents.

Brail bars (Figure 11) have been used by commercial mussel fishermen for many decades (Claassen 1994). They are dragged slowly by boat across the bottom of a large river (or lake; see Ortmann 1911); mussels clamp down on the brail hooks and are brought up onto the boat. Brails have been used by some investigators working in large rivers (which can be hard to sample by other means) to locate mussel beds and get some idea of their species composition (e.g., Starrett 1971; Nelson 1982; Isom and Gooch 1986; Williams and Schuster 1989). Nevertheless, brailing collects a small and highly biased sample of the mussel community and, therefore, is unsuitable for many kinds of surveys. Nelson (1982) described how to build and operate a brail bar.

Dredges (Figure 11) are another method used to collect bivalves by commercial fisherman, chiefly in marine waters. They have not been used widely to sample mussels in freshwater but probably have a potential similar to that of the brails to collect qualitative samples in some situations (e.g., deep waters) when other sampling methods are difficult to use (e.g., Negus 1966; Miller et al. 1989; Wilcox et al. 1993).

Impacts of Sampling Activities on Mussels

The sampling activity itself may have impacts on the mussel population that is being studied. Mussels may be trampled by investigators or may be disturbed by being handled. Some investigators collect mussels in a dive bag and then toss them back into the water at a convenient place, thereby potentially displacing mussels from suitable habitat. These activities may cause mussels to abort glochidia (e.g., Lefevre and Curtis 1910; Yeager and Neves 1986), reduce growth rates, or increase mortality rates. Little is known about whether and how sampling and other scientific activities affect mussels (but see Cope and Waller 1995; Dunn et al. 2000; Smith et al. 2000). Given the increasing intensity of sampling activities on some populations of rare mussels, it would seem desirable to carry out controlled studies on the effects of handling and trampling on mussels. Until such studies are done, it would seem prudent to disturb mussels as little as possible, handle them gently, and rebed them carefully in the sediment from which they were taken.

Summary

Mussels typically are collected by visual search or sediment excavation. Visual searches range from quick and superficial to slow, careful searches that detect all mussels at the sediment surface. However, visual searches do not detect buried mussels, which may be a large part of the population. Some amount of excavation and sieving sediment is needed to include the entire mussel population. Because excavation is expensive, we recommend double sampling to reduce the amount of excavation required. Mesh size is an important consideration when excavating to collect young mussels. Other methods for collection of mussels that are sometimes useful include grabs, dredges, tactile searches, muskrat middens, and brail bars.

Collecting mussels by any method might negatively impact reproductive behavior or survival. Until we know more about the effects of handling and trampling on mussels, it is best to disturb mussels as little as possible, handle them gently, and return them carefully to the sediment from which they were taken.

Making Observations on Mussels

Many attributes of a mussel community may be of interest to a surveyor. Most simply (and very commonly), you may be interested in whether a species is present in a specified area. A related variable commonly of interest is the species richness (number of species) in a specified area. This is equivalent to asking how many of the species in the regional species pool are present in a specified area and so is generally approached as is species presence–absence. Another variable commonly estimated is the size or density of a mussel population in a specified area. The density of one or all species in an area may be estimated. Finally, it may be desirable to measure other attributes of a mussel population, such as biomass, size structure, age structure, spatial distribution, sex ratio, reproductive status, body condition, parasite load, and so on. The variables that will be observed and recorded by the surveyor will depend on which of these attributes are of interest. For example, if species-specific density or abundance is of interest, then mussels must be identified and counted within each sampling unit; if survival is of interest, then mussels must be individually marked and marks recorded at each sampling occasion; and if spatial distribution is of interest, then the coordinates of sampling units must be recorded in addition to the count per sampling unit. It is worth emphasizing that a sound sampling design is just as essential for accurately estimating parameters such as sex ratio, parasite load, or physiological condition, as it is for estimating mussel population size (see also Payne et al. 1997).

Procedures for data management, quality assurance, and quality control should be developed before implementing the survey. Data management starts with a well laid-out data sheet and legible data recording. Although these might seem like obvious or trivial considerations, the cost of collecting survey data can be large and loss of data through poor data recording and management is inexcusable. A well thought-out database design including data entry forms can greatly ease the process of transferring data from field forms to computers and getting data ready for analysis. Well-designed data entry forms, using readily available database software, can also help assure accuracy through built-in error-checking functions. Data entry and error checking is time-consuming, and resources to do this job should be considered as the survey is planned and budgeted.

Accurate species identification is often critical to a successful mussel survey. Mussel identification can be challenging because of the subjective nature of some of the identifying characteristics. In the appendix, we provide advice and resources for mussel identification.

Summary

Many observations other than the number of mussels of each species might be made on freshwater mussels depending on survey goals and objectives. Demographic characteristics, such as mussel size, sex, or reproductive state, might be of interest if population dynamics or viability are being studied. Tissue samples might be taken if population genetics, bioaccumulation of contaminants, or physiological status are the focus of the study. Sound sampling design is important regardless of what variable is being observed.

Maintaining a system to assure accuracy in data recording and management is fundamental to a well-designed survey. Quality control and quality assurance measures should be implemented for the whole data management system, from design of field forms to data entry into computer format. Accurate species identification has obvious importance; we provide guidance on species identification in the appendix.

Data Analysis

Assessment of Precision and Power before Implementing a Survey

After you designed the survey, but before implementing it, you should consider carefully the precision or power of the proposed survey to see if it is adequate to meet the stated objectives of the survey. To begin this process, survey objectives must be stated in quantitative terms. If your objective is to estimate the number of purple lilliput *Toxolasma lividus* in a defined section of stream, consider how precise your estimate must be to be useful to you. Is it sufficient to know that there are between 100 and 1,000 animals, or do you need to know that the number of animals is between 150 and 180?

The object of survey design is to achieve the most precise estimate given the resources available to conduct the survey. Precision is determined chiefly by only two factors: density (or abundance) and survey design (especially sample size; e.g., number of quadrats). The importance of density arises from the empirical tight relationship between density and population variance (Downing and Downing 1992; Strayer et al. 1997).

Once the required precision is defined, you can estimate the number of samples that would be required to reach that precision. Using an estimate of the mean density of mussels from preliminary samples, the number of samples needed in a simple random design is

$$n = m^{-0.5}\mathrm{CV}^{-2},$$

where CV is the coefficient of variation (i.e., the standard error divided by the mean), m is the mean number of mussels per quadrat, and n is the number of quadrats searched (Downing and Downing 1992). The number of samples required to attain specified levels of precision given various mussel densities and simple random sampling is shown in Figure 13. (Remember that other designs may give much better precision than simple random sampling.) Downing and Downing's equation is based on samples taken from a relatively small, homogeneous area. Strayer et al. (1997), working on more heterogeneous sections of streams 100–200 m long, found that

$$n = 2.6m^{-0.51}\mathrm{CV}^{1.82},$$

and presumably, as the study area gets larger and more heterogeneous, the Downing and Downing regression will systematically underestimate the number of samples required to reach a given level of precision. Payne et al. (1997) also found that Downing and Downing's equation underestimated sampling requirements in large rivers, by about a factor of two.

When planning a survey that uses two-stage sampling, Seber (2001) showed how to calculate sample size for desired precision based on a pilot survey. For example, consider a river-wide survey where the river was partitioned into *N* reaches and a simple random sample of size n_0 reaches was selected for subsampling in a pilot survey. Abundance was estimated in each of the n_0 selected reaches, and variance among reaches was estimated by

$$s_0^2 = \sum_{i=1}^{n_0} \frac{(\hat{\tau}_i - \bar{\hat{\tau}})^2}{n_0},$$

where $\hat{\tau}_i$ was the abundance estimate from the *i*th reach and $\bar{\hat{\tau}}$ was the mean of n_0 abundance estimates. Then, the number of reaches that need to be sampled to achieve the desired precision (say, CV*) is

$$n = \left(\frac{M \cdot s_0}{\bar{\hat{\tau}} \cdot \mathrm{CV}^*}\right)^2.$$

Likewise, data from preliminary samples, empirical relationships (e.g., Downing and Downing 1992), or even intelligent guesses can be used to estimate precision of other sampling designs by using published formulas (e.g., Thompson 1992) for means and standard errors, given different numbers of projected samples.

It may be useful to adopt a graduated notion of acceptable precision, allowing for less precision as density decreases. As density decreases, higher sample sizes are needed and expectation for precision should be relaxed. In general, for probability-based surveys using 0.25-m^2 quadrats and an efficient sampling method, such as systematic sampling, highly precise estimates (CV < 0.20) can be achieved with a sample size of 300 quadrats if density is greater than or equal to 0.50/m^2 (Smith et al. 2000). Precise estimates (CV < 0.30) can be achieved with a sample size of 500 quadrats if density is greater than or equal to 0.20/m^2. Moderately precise estimates (CV < 0.35) can be achieved with a sample size of 700 quadrats if density is greater than or equal to 0.10/m^2.

Even if density for the targeted species is less than 0.10/m^2 and estimates are likely to be imprecise (Figure 13), probability-based sampling may be worthwhile when the objective is to estimate density or abundance. First, without probability-based sampling, there is no measure of the reliability of the estimate. Second,

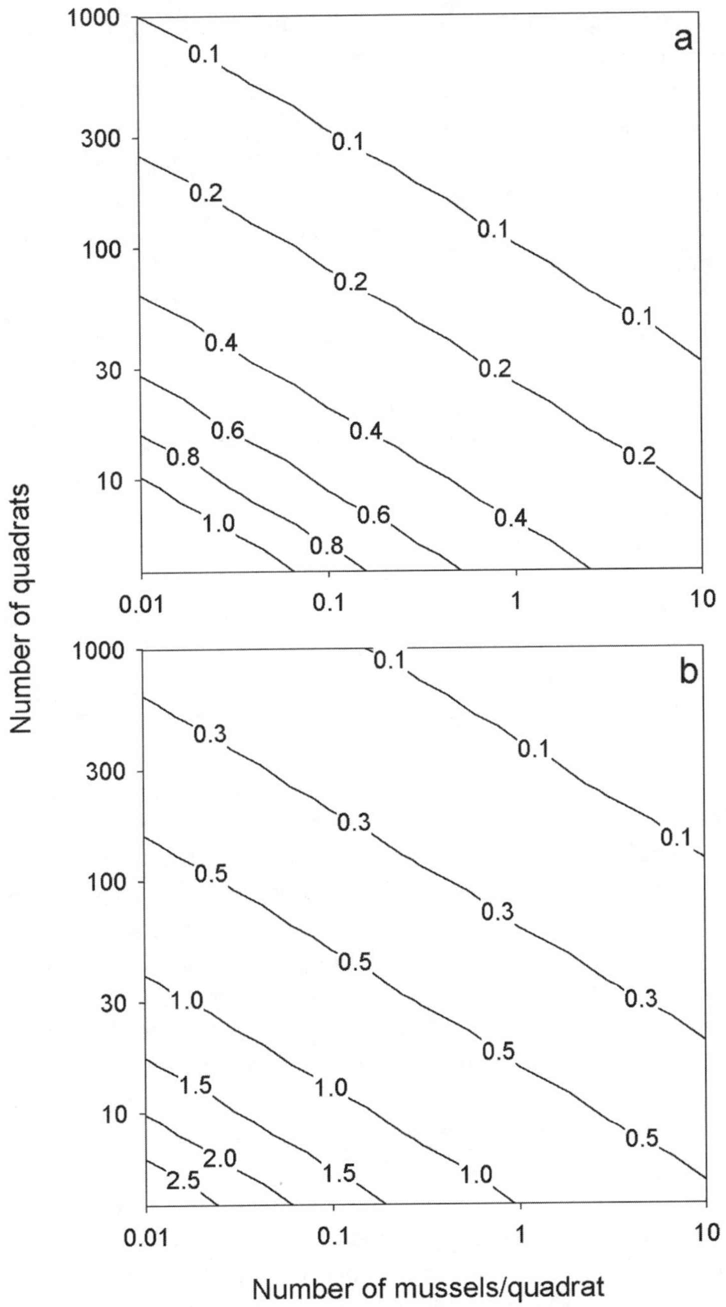

FIGURE 13. Precision of estimates of mussel population size or density as a function of mean number of mussels/quadrat and number of quadrats sampled using a simple random sample design. Panel (a) shows expected coefficient of variation and panel (b) shows half the width of the 95% confidence interval (i.e., $t \times$ SE). Based on Downing and Downing (1992). Remember that other sampling designs may give much better precision than simple random sampling.

although the target species may be sparse, populations of other species at the site may be dense enough to be estimated with adequate precision. Third, if density of the targeted species increases at the site, then probability-based sampling is the best approach to detect the recovery by separating a true increase in density from sampling error. Fourth, application of probability-based sampling allows for a probability statement to be made about species absence and maximum density at a site even if no mussels are found (Green and Young 1993).

If the estimated precision of the sampling program is insufficient to meet the objectives of the survey, then the objectives must be rethought, the target population must be redefined, a more efficient sampling design must be used, or available resources must be increased to allow for greater sampling effort (Figure 1B). After changes in objective, design, or resources are made, the survey must be reevaluated. If no affordable survey designs meet meaningful objectives, then the survey effort should be abandoned. There is little point in conducting a survey that cannot adequately address your objectives.

Statistical power is the ability to reject a null hypothesis when it is in fact false. If you are conducting a mussel survey to test a hypothesis (e.g., that mussel populations have changed since an earlier survey or that mussel populations are lower downstream of a pollutant discharge than upstream), then it is worth considering whether the proposed survey design will have adequate statistical power. Many ecological studies have low statistical power (e.g., Steidl et al. 1997); just as there is little point in conducting a survey if the results will be too imprecise to address your objective, there is little point in doing a survey if the resulting data are too weak to adequately test your hypothesis. Methods for estimating statistical power are varied, depending on the statistical test being used, but basically involve specifying an effect size (the minimum difference between the treatments that the investigator would like to detect), the α-value (conventionally set at 0.05, although, increasingly, it has become apparent that ecologists sometimes should use more liberal α-values; Osenberg et al. 1994; Steidl et al. 1997) at which the hypothesis will be tested, and the expected characteristics of the data (number of samples, expected variance). A fine source of information about statistical power, including a bibliography and links to sites with free programs for calculating statistical power, is *http://www.im.nbs.gov/powcase/powlinks.html.* Many general statistical packages and specialized statistical packages now calculate power as well—see Thomas and Krebs (1997) for a review of some software. Other good references include Cohen (1988) and Peterman (1990).

An example of statistical power calculations for a simple random sampling design is shown in Figure 14; note that statistical power is inadequate in this case for many combinations of mussel density, quadrat number, and population change. Although it may seem inconvenient to go through this exercise before starting your survey, it will be much more inconvenient to discover after the survey that the data set is too weak to effectively test your hypothesis. The solution to inadequate statistical power is to redesign the survey, dedicate more resources to the study, be satisfied with a larger effect size or a larger value of α, or abandon the proposed work.

The precision of presence–absence data presents a special problem. Conservation biologists used to working with large vascular plants or even vertebrates may think of presence–absence as being a robust or even absolute variable that can be estimated without error; but, for cryptic animals like mussels, presence–absence rarely can be established absolutely, except for small study areas. Instead, stating that a mussel species is "present" is equivalent to saying that mussel population

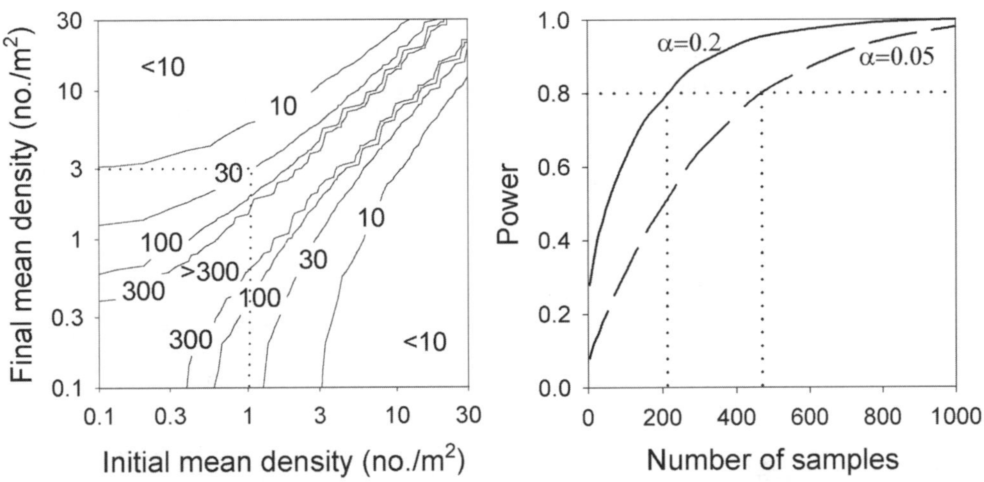

FIGURE 14. Examples of power analyses. The left-hand panel shows the number of 0.25-m^2 quadrats that must be sampled using a simple random design to find a significant difference (with a two-tailed *t*-test) between two populations with given mean densities. For example, if we wished to detect a difference between populations with densities of 1/m^2 and 3/m^2, we would need to take about 30 samples from each population (dotted lines). For this example, we specified that $\alpha = (1 - \beta) = 0.15$; that is, we were willing to accept a 15% probability of false positives (concluding that a difference exists when the means are really equal) and false negatives (concluding that no difference exists when the specified difference really exists). If we desired lower rates of false positives or false negatives, we would need to take more samples.

In the right-hand panel, we analyze a specific problem: our ability to detect a 30% decline in mussel population density from a starting density of 1/m^2. The graph shows the power (i.e., the probability of detecting a statistically significant difference using a one-tailed *t*-test) as a function of the number of 0.25-m^2 quadrats sampled with a simple random design. The two curves are based on a conventional value of $\alpha = 0.05$, as well as a more relaxed value of $\alpha = 0.2$. If we desired to have a power of 80% to detect the decline, we would need to sample 467 quadrats for $\alpha = 0.05$ and 214 quadrats for $\alpha = 0.2$. Samples taken from designs other than simple random designs may have much higher power and lower sampling requirements than these examples, whereas samples taken in heterogeneous sites may have lower power and higher sampling requirements.

density is above some detection threshold and can be estimated only with error. To a first approximation, the probability of detecting at least one individual in a rare mussel population (i.e., $<0.1/m^2$) is

$$p(\text{detection}) = 1 - e^{-mn},$$

where m is the mean number of animals in a sample, n is the number of samples, and e is the base of natural logarithms (Green and Young 1993). It is straightforward to use this formula to estimate the probability of detecting a population, given a target density and assuming a number of quadrats or transects (Figure 15). Green and Young's formulas have been generalized for a double sampling design (Smith et al. 2001).

It is a little more complicated to estimate the probability of detecting a mussel population using timed searches. Following Green and Young (1993),

$$p(\text{detection}) = 1 - e^{-R},$$

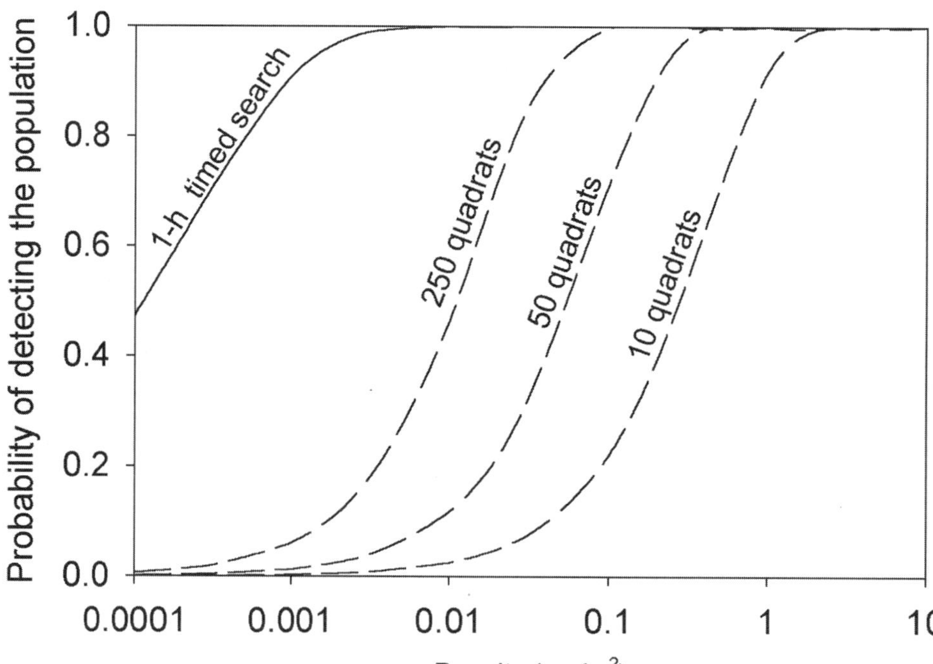

FIGURE 15. Plot of probability of detecting a mussel species as a function of population density and effort. Estimates for quadrats from Green and Young (1993); estimates for timed searches from Strayer et al. (1997) and technically apply only to Strayer searching for eastern elliptio. The x-axis is logarithmically scaled.

where R is the mean number of animals detected in a timed search, that is, the length of the search times the encounter rate (Strayer et al. 1996; Strayer 1999a). The encounter rate is the rate at which individual mussels are found and is commonly reported as catch/h. It is not straightforward to relate encounter rates in timed searches to actual population densities—this relationship presumably varies greatly across mussel species, investigators, and field conditions. Further, in the only case where this relationship has been investigated, it had very large scatter (Strayer et al. 1997). Nevertheless, it is clear that the probability of detecting a mussel population in a timed search depends on the encounter rate and the length of time spent searching and is usually an unknown function of mussel population density. An example is shown in Figure 15. As Metcalfe-Smith et al. (2000) and Lellis (2001) have shown, mussel populations may be so sparse and cryptic that they are detected only with long (>5 person-hours) timed searches (Figure 16). Other useful references from the growing literature on statistical approaches to the detection of rare species include McArdle (1990), Nicholson and Barry (1995), and Venette et al. (2002).

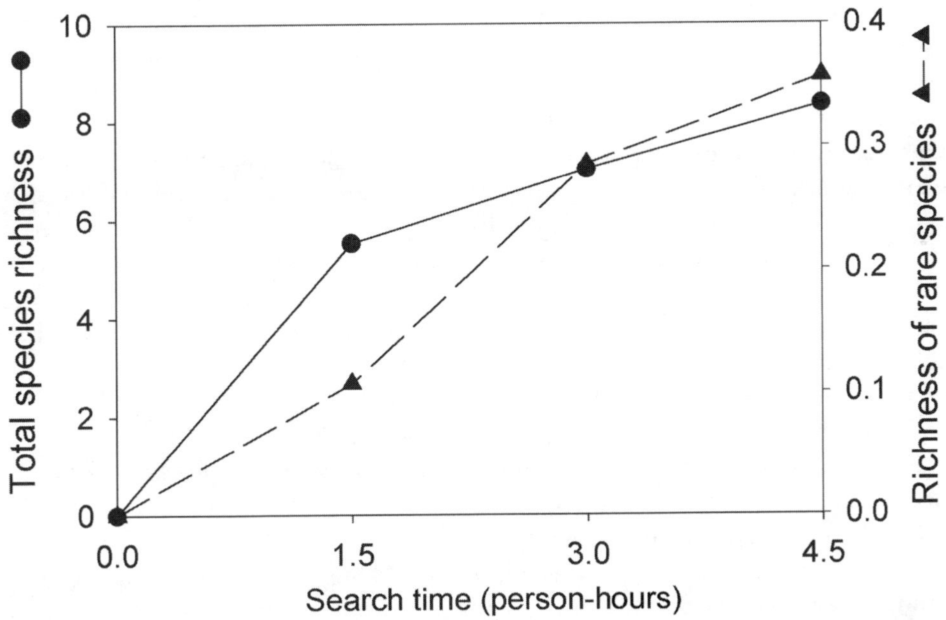

FIGURE 16. Example of increase in species richness with increase in effort, showing that very great effort may be required to detect all mussel species present at a site. Solid line shows all species, and dashed line shows the three rare species (rayed bean *Villosa fabalis*, northern riffleshell *Epioblasma torulosa rangiana*, and wavyrayed lampmussel *Lampsilis fasciola*) that were special targets of the surveys. Note the difference in scaling between the two *y*-axes. From Metcalfe-Smith et al. (2000).

Because mussel presence–absence depends so strongly on survey design, it is once again important to evaluate whether a proposed presence–absence survey is really capable of detecting the mussel population that is the objective of the survey. If not, the survey should be modified (or abandoned) or the objectives should be rethought.

Analysis of Observations from a Completed Survey

Interpretation of survey results depends critically on the design of the survey (Table 1). Results from the designs that we have discussed range from species lists to rigorous estimates of total abundance. To avoid reaching incorrect conclusions, it is important to keep in mind how observations were made before analyzing the observations and interpreting the results.

Although infrequently applied, complete coverage of a site is an option in some mussel surveys. In the event that a site is completely covered, the collection method may involve a superficial or detailed search. Because the site is completely covered, there is no sampling error to estimate; however, the number of mussels counted is subject to error because not all mussels will be detected. Unfortunately, we cannot adjust for the imperfect count or measure the error around the count because the proportion of mussels detected is unknown. In the case of a superficial search, the result is an index of population abundance, and the proportion detected (i.e., the ratio of the index divided by the true population abundance) is, in general terms, a product of two quantities: (1) the probability of detection given that a mussel is at the surface and (2) the proportion of mussels at the substrate surface. Variation in the probability of detection or the proportion at the surface could cause misleading comparisons across place, time, or taxa. In the event that a detailed search is used in combination with complete coverage, then we can assume that all the mussels at the surface are indeed detected and the survey result can be interpreted, not as total abundance but as the abundance of mussels at the substrate surface. However, here again, comparing the result across place, time, or taxa could be misleading because the proportion at the surface tends to vary widely.

The main shortcoming of informal sampling of habitat within a site is the lack of a valid method for assessing sampling error even though results are affected by substantial uncertainty because the entire site is not covered. In addition, because informal sampling is usually coupled with collecting mussels by a superficial search, the proportion of mussels detected is unknown. Thus, abundance is biased, and a standard error or confidence interval for an estimate of abundance cannot be calculated. Also, in a comparison between counts across place, time, or taxa, there is no way to decide whether an observed difference is a result of sampling error, variation in the proportion detected, or a difference in true abundance. Thus, when the survey design involves informal sampling, interpretation of the results should be limited to a list of species present. The completeness of that list depends on spatial coverage and effort.

The important advantage of probability-based sampling is that sampling error can be quantified and used to gauge the reliability of an estimate. Accounting for sampling error is integral to rigorous statistical inference and allows uncertainty to be incorporated into decision making. Sampling error, as measured by standard error, is used to compute confidence intervals and to test whether differences in estimates are real or due to chance alone. Probability-based sampling can be coupled with detailed search, excavation, or a combination. If excavation is not included, then the survey results apply only to mussels at the substrate surface and are subject to confounding due to variation in the proportion at the surface. The only survey design that will result in rigorous estimates of total abundance is probability-based sampling with some amount of excavation.

The procedures or formulas used to analyze observations and arrive at results also depend on survey design. With complete coverage or informal sampling, analysis simply entails taking sums or means of counts, examining length distributions through frequency histograms, and other descriptive procedures. For probability-based sampling, analysis is specific and can be complex.

Because analysis can be complex for some of the more useful designs, user-friendly software is welcome. Software for estimating species-specific density and abundance based on the systematic and double sampling designs presented in Smith et al. (2000) can be accessed at *http://www.lsc.usgs.gov/aeb/2068/index.asp.* Software for surveys that use adaptive cluster sampling can be accessed at *http://www.lsc.usgs.gov/aeb/davids/acs/index.asp.*

Summary

We advocate a step-by-step framework to survey design that begins with a quantitative statement of survey goals, requires a clearly defined target population, moves to a statistically valid sampling design, and includes an iterative step where you ask whether the survey as designed is sufficient to meet the intended goals and make adjustments if necessary. The last step occurs before a survey is implemented and is analytical in nature, using sample-size calculations and power analysis. Including these analyses prior to survey implementation allows you to adjust the goals, target population, or sampling design if you find that the current survey will not meet the goals. Often it boils down to increasing sampling effort, and sometimes resources are not available to do that. However, we feel strongly that there is little justification for implementing a survey that is insufficient to meet its intended goals. Even when there is administrative or regulatory pressures to "just do something," results from sample-size calculations or power analyses can be used to argue for sufficient resources to do the job right.

The final steps in a survey are to analyze the observations and draw conclusions from the analyses. The analytical procedures will depend on the underlying sampling design, the data type (e.g., continuous or categorical), and the objective (e.g., estimate spatial distribution, estimate density, or estimate population change). Much statistical analysis involves the quantification of uncertainty, and, when sampling freshwater mussels, there are two important sources of uncertainty. One source of uncertainty comes from the sampling process where mussels are collected from a sample of the study site. Another source of uncertainty comes from imperfect detection of mussels within the sampled area. Proper sampling design and analysis accounts for both sources of uncertainty, so that conclusions are valid.

Sampling Approaches for Some Common Objectives

Preliminary Surveys

By "preliminary survey," we mean collecting information about a mussel population or sampling method before beginning the formal survey. Because good survey design is based on knowledge of the target population and site characteristics, a preliminary survey of a poorly known site will almost always improve survey design, often substantially.

Preliminary surveys vary widely in design and cost. We stress that even the briefest and simplest of preliminary surveys may yield valuable information and that the more thought and effort is put into a preliminary survey, the more useful it can be. If you are planning to survey the mussel populations of a wadable stream you have not seen before, an hour cruising the site will often give you an idea of the density, distribution, and species composition of mussels to be expected, and will allow you to design a more effective survey than you could sitting in your office. In larger rivers, you may be able to use brail bars, grabs, dredges, or hydroacoustic methods to locate mussel beds and determine the kind of sediment you will be sampling. Local anglers and outdoor enthusiasts may be able to provide useful information on the location of mussel beds (although we have been surprised at how often such people have not been able to provide accurate information).

More formal preliminary surveys often are helpful as well. Suppose we decide from our 1-h walk through a wadable stream that most mussels are clustered along the banks. That same day, we could do a quick stratified random survey with 0.25-m^2 quadrats, searching 25 quadrats along the banks and 25 in midchannel, and recording the time it takes us to do the work. When we return to the office to design the formal survey, we have actual (although rough) estimates of mussel densities and variances in the two strata, along with an estimate of the time needed to search each quadrat. This information can be used to assess the precision and power of various designs, estimate the cost of the survey, and make optimal allocations of sampling effort to the two strata.

Preliminary surveys are useful in designing multiple-site surveys as well. Double sampling for stratification (Box 4) is an example of how inexpensive preliminary surveys can be combined with later, more intensive work to provide good estimates of mussel populations over large regions.

If the study area is large and complex, and the formal survey will be expensive and time-consuming, it may be worthwhile to conduct extensive preliminary work, perhaps even devoting a field season to preliminary work before attempting the formal survey. Although there is often pressure to finish survey work quickly, time spent in reconnaissance and preliminary sampling is often well spent.

Determining Species Presence or Species Richness at a Single Site

Two of the most common objectives of mussel surveys are to determine if a species is present at a site and to determine the list of mussel species that live at a site. If the site is very small, it might be completely covered; otherwise, the site must be sampled in some fashion. Designs to meet these objectives should have a high and known probability of detecting mussel species. Unfortunately, designs that have a high probability of species detection usually do not allow investigators to estimate that detection probability, and designs that readily allow estimation of detection probabilities usually do not have high detection probabilities.

Traditionally, surveys to detect species presence or develop lists of mussel species living at a site have been informal designs using superficial visual surveys. That is, investigators range widely over the study site, picking up mussels as quickly as possible. Such surveys offer very good detection of mussel species (Figure 15; see also Hornbach and Deneka 1996; Strayer et al. 1997; Vaughn et al. 1997; Obermeyer 1998). Excavation is too slow a technique to be useful in species detection, unless species that bury deeply in the sediment (e.g., clubshell) are especially being sought. Mask-and-snorkel and diving searches probably are more effective than wading searches, especially for small or cryptic mussels (e.g., Strayer et al. 1997). Because mussels often occur in mixed-species beds, it makes sense to concentrate effort in mussel beds, if they are found. Nevertheless, it is important that the search achieve good spatial coverage of the site even if it means searching areas that might not be thought of as "good mussel habitat." It is especially important not to overlook special habitats (e.g., backwaters, banks), which may harbor species not found elsewhere at the site. Searching along the shores for spent shells and muskrat middens often is a quick way to develop or add to a species list for a site.

Unfortunately, the inability to estimate the detection probabilities of informal designs seriously limits their use in determining the presence of mussel species. Except for the rare cases in which the entire study area can be completely searched, it is not possible to say with certainty that a mussel species is absent from an area.

As we have noted elsewhere (page 57), our ability to find a mussel species depends on the sampling design and effort. So instead of being able to say that a species is truly absent from a site, we must content ourselves with saying that the species was not detected given *X* effort and *Y* design, or that a design with a 80% chance of detecting a population with a density of $0.05/m^2$ failed to find the species. It is not usually possible to say anything about the range of possible population densities for a population that was undetected by an informal search (but for an exception see Strayer et al. 1996, 1997), and problems with the repeatability of informal surveys (page 78) mean that a statement like "we failed to find a clubshell in a 2-h timed search" is of limited value. The fact that there is an unknown chance that rare species could be missed by an informal search is, therefore, unacceptable for many goals (e.g., determining the presence–absence of an endangered species at a construction site or rigorously assessing temporal change).

Instead, in many settings (especially if the purpose of the survey is regulatory in nature), it will be necessary to use a design for which you can estimate the probability of detecting a mussel population of a certain size or density. Here, the challenge is to design a survey that reduces the chance of missing the presence of a species to an acceptable level. The chance of missing a species that is actually present at the site (equivalent to a type II error) decreases with increased species density and with increased sampling effort and spatial coverage (e.g., Figures 15 and 16).

Green and Young (1993) presented a theoretical basis for controlling the chance of missed presence, an approach that has been implemented for assessment of endangered species (Strayer et al. 1996; Smith et al. 2001). If the counts of a species in quadrats or transects are Poisson distributed (as is often the case for rare species), then the number of quadrats or transects that must be sampled is

$$n = \frac{-\ln(\beta_0)}{\bar{a}m},$$

where β_0 is the acceptable chance of a missed presence, $\bar{a}$ is the area of the sampling unit (in m^2), and m is the density (per m^2) of the target species. If the only objective of the survey is to detect a particular species, searching could stop as soon as the species was encountered, thus reducing the cost of the survey, especially at sites where the species was abundant. Venette et al. (2002) described other useful designs for detecting rare species and estimating the probability of missing rare species when counts are not Poisson distributed.

Because of the expense of implementing quantitative sampling, semiquantitative sampling designs (Table 1) may be appropriate for surveys designed to determine whether a particular species is present. For example, 1-m-wide transects oriented across a river (e.g., bank-to-bank) and located systematically with random starts throughout the site would provide the basis for a good design. If the transects were searched visually or tactilely, some individuals within a transect would go undetected so that the probability of detection within the area sampled would still be unknown. However, the probability of detection could be estimated by periodically locating quadrats along the transects for sediment extraction. The Green and Young formula could be adapted for semiquantitative sampling by incorporating a parameter for the probability of detection. Thus, the necessary sample size could be calculated by

$$n = \frac{-\ln(\beta_0)}{\bar{a} \cdot \lambda \cdot m},$$

where β_0 is the acceptable chance of missed presence, $\bar{a}$ is the average area of the sampling unit (e.g., for 1-m, cross-river transects, $\bar{a}$ is equal to the average transect length, in m), λ is the probability of detection during a visual/tactile search, and m is the density (per m^2) of the target species. However, we are unsure if the above formula holds when the sampling unit is large (e.g., cross-river transects) because of the underlying Poisson assumption. We strongly recommend field trials of the survey protocol to test its efficacy.

The importance of a rigorous survey design for determining species presence is apparent when considering endangered species assessment. Suppose an informal and untested survey protocol was implemented to determine presence of an endangered species at the site of a potential impact and the endangered species was not detected. The finding of species absence could be challenged because of the ambiguity of "species absence" and the inadequacy of an informal and untested survey protocol. It could be argued that the finding of absence from an informal and untested survey protocol has no meaning because the chance of missing a species that was present at the site was unknown and, therefore, could be unacceptably high. It would be difficult to construct an objective, science-based counterargument to that challenge.

It may be useful to combine informal designs with formal designs. For example, a survey designed to detect a protected species could begin with informal searches. Formal surveys could be added at sites where the species was not detected by the informal searches, thus allowing the investigator to calculate the chance of a missed presence and the maximum probable density of the missed species (e.g., Smith et al. 2001).

Determining Species Presence–Absence across Multiple Sites

An investigator may wish to know about the presence or absence of species across a broad geographic area, such as a drainage basin or state. While methods for surveying a site are the same as those just described, sites to be surveyed should be chosen carefully to match the objectives of the survey. There are perhaps two specific questions that motivate surveys of this sort: to establish the total list of species in an area and to establish the range of individual species in an area.

If the objective is to establish a species list for an area, it probably makes most sense to begin by surveying the most diverse sites in the region, then add distinctive sites that are likely to contain additional species. As a rule, mussel species richness rises with stream size (van der Schalie 1938; Strayer 1983; Watters 1992), so the larger streams and rivers in the study area should be examined carefully. Remaining habitats in the study area can be subsampled for study, concentrating on special habitats (e.g., lake outlets) or sites that are known to have high water quality or diverse communities of fish (Watters 1992) or invertebrates. MacKenzie et al. (2002) presented methods based on mark–recapture sampling to estimate site-occupancy rates, a variation on the species richness objective, and Bayley and Peterson (2001) took a Bayesian modeling approach to estimating the probability of species occurrence in multisite surveys of fish species.

Bridge crossings are traditionally foci of mussel surveys, but higher quality habitats may be found away from bridges, so float surveys often are a useful tool (e.g., Vaughn and Taylor 1999; Lellis 2001). Because bridges may have been built at historic fords (i.e., riffles) or bridge construction may have had an impact on local mussel populations, bridge crossings may not be representative of a river or a watershed. Remember that bridge construction could have affected mussels and their habitat upstream, downstream, and directly beneath the bridge.

Although malacologists often speak about the range of mussel species in general terms like "the Huron River between Dexter and Ann Arbor," or present spot maps that show known occurrences of a species, procedures for defining the range of a mussel species have not been standardized. Two quantities must be specified to define the range of a species: the grain size at which the range is defined and the minimum density within each grain that is required for the species to be considered "present" in that area. (We are using "grain size" here in the sense of landscape ecology to mean the finest level of spatial resolution in the data set [Turner et al. 2001], not to refer to the particle size of sediments.) The grain size sets the level of resolution at which the range will be defined. Grain size may be very coarse, as, for example, in the case where mussel ranges are given in terms of the states where the species occurs (e.g., Williams et al. 1993). Within a stream system, it probably is most natural to set the grain size as a length or area of stream bottom. For example, Lellis (2001) divided a river into 200-m-long segments in their mussel survey of the Delaware River, and Strayer (1983) divided streams into 8.1-km-long (5-m-long) segments in a study of mussel distribution in southeastern Michigan.

Other possibilities exist as well; a stream network may be divided into links (segments between the entry of tributaries). Finer grain sizes will give better resolution of range but may require substantially more work. Once the grain size is set, the investigator will need to set a threshold density below which the species is considered to be absent from a site. As we have already noted, it rarely will be possible to do a complete census of the sites or be certain that a mussel species actually is absent (as opposed to rare) from a site. This threshold could be set in quantitative terms (e.g., species *X* is defined as present if more than *T* individuals exist in a 1-km reach or if the mean density exceeds D/m^2) if quantitative sampling methods are used. However, search rates using quantitative methods are so inefficient at detecting rare species (Figure 15) that such methods usually will not be suitable for defining species ranges. Instead, it probably is best to use timed searches to establish species ranges and to set the detection threshold in terms of minimum encounter rates (e.g., species *X* is defined as absent if no live animal is detected in *Y* person-hours of searching). Because the relationship between actual mussel population densities and catch rates from timed searches appears to be very variable (Figure 17), ranges so defined should be interpreted cautiously, especially when catch rates are near the

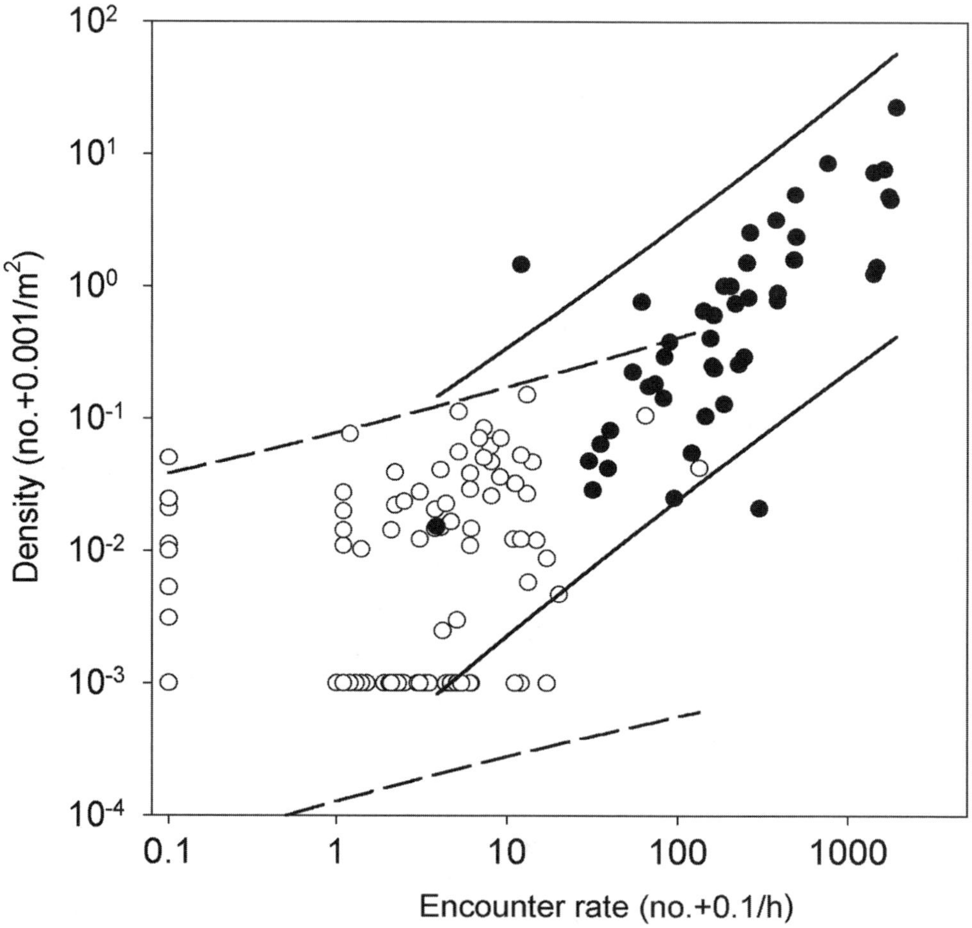

FIGURE 17. Relationship between catch-per-unit effort (CPUE) in timed searches, while snorkeling (by D. L. Strayer), and population density in streams of the Atlantic Slope (modified from Strayer et al. 1997). Black dots show data for eastern elliptio, white dots show data for all other species, and curved lines show 95% prediction intervals on individual points. Note that these intervals span ranges of 100–1,000-fold, so that CPUE is a very poor estimate of actual population density.

detection threshold. An example showing the sensitivity of range size to grain size and detection threshold is shown in Box 9.

Determining Population Density or Size within a Defined Area

This is a very common objective, either as an end in itself, or as a means to address objectives about impacts or temporal change (see below). If the study area is small, then the survey design generally will be straightforward. If the mussel population is relatively uniform (as in a large dense mussel bed), a simple random design may suffice. If the mussel population is patchy, it often will be worthwhile to do some preliminary sampling to locate the densest populations so that a stratified design can be used, or use systematic sampling, which is efficient for rare and clustered populations. Alternatively, double-sampling (or possibly adaptive) designs may be used. In general, these more sophisticated designs will give more precise results than a purely random design and are only a little more complicated. The choice of sampling and collection methods will depend on the target

BOX 9. Defining the Range of a Freshwater Mussel

Lellis (2001) conducted an extraordinarily detailed survey of the middle Delaware River along the New York–Pennsylvania border that nicely illustrates the issues that need to be considered in defining range size. Lellis and his coworkers snorkeled an entire 118-km reach of the Delaware, using timed searches to assess mussel populations in 595 contiguous 200-m segments. It is very unusual (probably unique) to have such spatially complete information on mussel populations.

Figure 18 shows the range of the creeper *Strophitus undulatus* in the middle Delaware defined according to different grain sizes and minimum densities. Figures 18a and 18b show the actual data collected by Lellis, suggesting an extensive but fragmented range (to the extent that the range can be discerned). In Figure 18c, the range is replotted using a minimum detectable density of 1 animal/h (roughly a standard in timed searches; see Metcalfe-Smith et al. 2000). Not surprising, the perceived range becomes much more sparse. Changing to a coarser grain size (Figure 18d) makes the range appear to be continuous over the entire study area. Finally, in Figure 18e, we show that range size defined using a coarse grain size and a high minimum detectable density suggests that the creeper does not occur in the study area (which was, in fact, the conclusion reached in an earlier, more superficial study by Strayer and Ralley 1991)! Which (if any) of the ranges plotted in Figure 18 is the "true" range of the creeper in the middle Delaware? There is no "true" range independent of the definitions of minimum detectable density and grain size adopted by the investigator. As Figure 18 shows, it is important to be aware of the importance of such definitions when discussing current or historic changes in range size.

(Box continues on the following page.)

BOX 9 (continued)

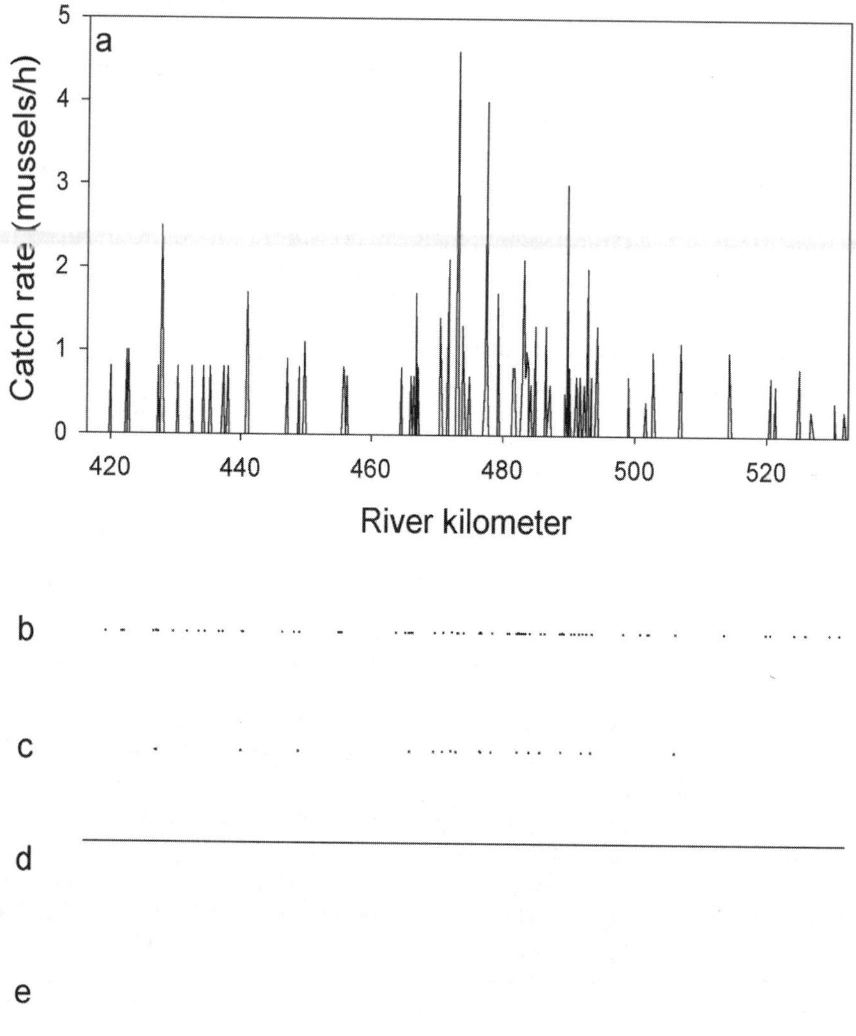

FIGURE 18. Range of the creeper in the middle Delaware River, defined by using different grain sizes and minimum detectable densities: (A) Actual catch rates of the creeper reported by Lellis (2001), at a grain size of 200 m and a minimum detectable density of ~0.7 mussels/h; (B) Range of the creeper defined using the data in (A); (C) Range of the creeper defined using a grain size of 200 m and a minimum detectable density of 1 mussel/h; (D) Range of the creeper defined using a grain size of 10 km and a minimum detectable density of 0.01 mussel/h (i.e., the range covers the entire study area); (E) Range of the creeper defined using a grain size of 10 km and a minimum detectable density of 1 mussel/h (i.e., none of the river is within the species' range).

population (especially if juveniles are to be included), the characteristics of the study site, and available resources. If juveniles are sought or an estimate of total abundance is required, then it will be necessary to do some excavating and sieving, probably using a double-sampling design (Box 5).

If the study area is large and very heterogeneous (e.g., an entire drainage basin), then the sampling design may be complex. Commonly, double sampling with stratification at the first stage will be suitable in such cases (Box 4).

Regardless of the size of the study area and the study design, it is important to compare the expected precision of the survey results with the objectives. Estimates of mussel population density or size often are very imprecise (Figure 13) and may not be adequate to address the objectives of the study. Methods for estimating the precision of some simple designs are given in the section on precision and power.

Although timed searches often are used as a rough measure of population density, they offer no information about the actual size or density of a mussel population, and they are burdened with large and unknown errors (e.g., Figures 17 and 19). Hence, investigators should not use timed searches to assess the size or density of mussel populations without giving careful thought to the serious shortcomings of this approach.

Comparing Mussel Populations at Two or More Sites

A comparison of mussel populations across sites follows closely from the previous section. Population densities or sizes are estimated within each of the study areas and compared using any of several approaches (which are nicely summarized in Table 17 of Elliott 1977). Note that it is not necessary to use the same methods or sampling designs at all of the sites, but if different methods are used across sites, the investigator must be careful not to introduce bias from the use of different methods. Data may need to be transformed prior to analysis to meet assumptions of normality (Box 10). Most commonly, standard statistical techniques such as *t*-tests or analysis of variance (ANOVA) are used to test the null hypothesis that there is no difference among sites. Because estimates of population density or size within a site typically are imprecise, statistical tests often have low power to detect differences across sites (Figure 14). Consequently, it is again prudent to estimate the statistical power of a proposed sampling design to detect a difference across sites before embarking on the sampling program. If the sampling program is too weak to meet the objectives of the study, either the design or the objectives should be modified or discarded.

There is a growing feeling among ecologists that statistical significance tests are not useful for comparing populations or studying change in a population over time (Johnson 1999; Anderson et al. 2000). Significance tests lead to simple conclusions, such as density at site A is different than at site B. Such a conclusion may not be informative for several reasons. First, there is no information on the

BOX 10. Data Transformations

Parametric statistical analyses such as *t*-tests, ANOVA, and regressions assume that the data are normally distributed and that the variance is independent of the mean. Many ecological data violate these assumptions. If parametric statistics are used on data that do not meet these assumptions, the resulting *p*-values of significance may be incorrect.

The best transformation to use depends on the structure of the data. Several transformations are commonly used on the raw data before subjecting them to parametric statistical analysis (Table 3). If the data are counts that approximate a Poisson distribution, then a square-root transformation is commonly recommended; if the data set includes many small or zero counts, a $\sqrt{(X+1)}$ or $\sqrt{(X+0.5)}$ transformation may be more effective (Snedecor and Cochran 1967; Elliott 1977). Downing (1979) found that a 4th-root transformation was more effective than a square-root transformation for highly aggregated benthic data. For continuous data, a logarithmic transformation often is useful. If the continuous data set contains zeroes (for which the logarithm is undefined), then it is customary to use a log (X + 1)-transformation. The base of the logarithm is irrelevant. Several authors have pointed out that the addition of the constant 1 is arbitrary and may lead to important bias in some applications (McArdle et al. 1990), so constants other than 1 occasionally are recommended (Berry 1987; Yamamura 1999). If the data are proportions, then useful transformations are arcsin $\sqrt{X}$ and arcsin X, depending on if the data are counts or continuous, respectively.

TABLE 3. Some useful transformations for ecological data (Snedecor and Cochran 1967; Elliott 1977; Downing 1979; Downing and Downing 1992).

Situation	Transformation
Counts, Poisson-distributed (i.e., $s^2 \approx \bar{x}$)	$X' = \sqrt{X}$; $X' = \sqrt{(X+1)}$; $X' = \sqrt{(X+0.5)}$
Counts, $s^2 > \bar{x}$	$X' = \sqrt[4]{X}$
Continuous data	$X' = \log(X)$; $X' = \log(X+1)$
Counts as proportions	$X' = \arcsin(X)$; $X' = \arcsin\sqrt{X}$
Continuous data as proportions	$X' = \arcsin(X)$

After a transformation is applied (but before running a statistical test), it is wise to check that the transformation was adequate. Probably the simplest way to do this is to plot the mean against the variance of the transformed data for your samples; there should be no relationship between the mean and the variance. If the variance and mean still are related, you should try other transformations or use nonparametric statistical tests. Generally, transformations that eliminate the relationship between variance and mean also make the data close enough to normal to satisfy the needs of parametric statistical tests.

magnitude of difference or the range of plausible differences. Second, ecological significance and statistical significance do not necessarily coincide. For example, the observed difference might be statistically significant but not ecologically interesting. Conversely, the observed difference might not be significantly different from zero but might be ecologically important. Third, the null hypothesis of no difference often is not sensible in ecological studies. It is usually trivially true that any two sites have different densities of freshwater mussels. The relevant question is not whether the densities are different but rather how large is the difference.

As an alternative, the difference or change, the variance of the difference or change, and the confidence interval of the difference or change may be calculated instead of (or at least in addition to) significance tests (Anderson et al. 2000). A confidence interval is more informative than a significance test (Figure 19). Inference from a significance test is limited to "reject the null" or "fail to reject the null." In contrast, a confidence interval shows plausible values for population change or for the difference among populations—depending on what is being estimated. A confidence interval can be used to infer when a biologically significant difference or change occurred and when uncertainty in the estimate prevented a conclusion from being made—perhaps due to insufficient sample size or study design (Figure 19).

Detecting an Impact on Mussel Populations

Very commonly, investigators ask whether an activity (usually a human activity) has an impact on a mussel population. The most common designs in such situations are the so-called before-after-control-impact (BACI) design and its variants,

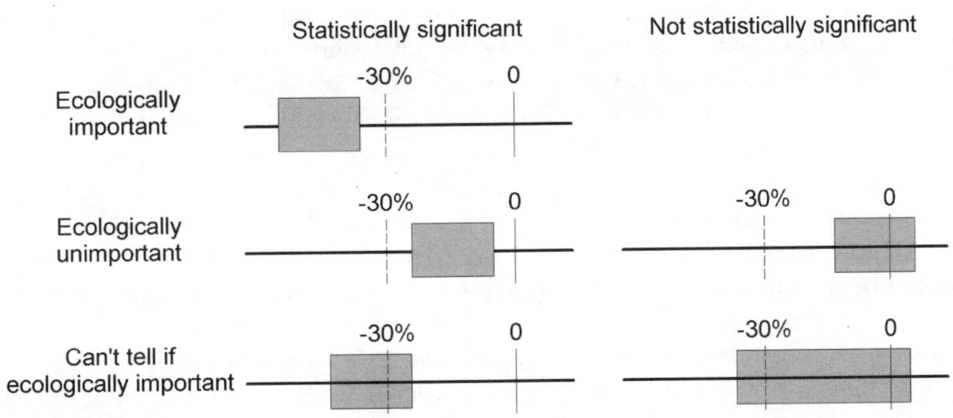

FIGURE 19. An illustration of the value of presenting confidence intervals, as well as the results of a significance test when comparing mussel populations across sites or times. Here, the gray bar shows the confidence interval at the value chosen by the investigator (e.g., a 90% confidence interval). The vertical bar at –30% represents the degree of change deemed by the investigator to be important. If the confidence interval fails to overlap 0, the difference is statistically significant (when the null hypothesis is that the true difference is 0). If the confidence interval lies entirely to the left of –30%, the difference is ecologically important. Note that nearly every possible combination of statistical significance and ecological importance is possible—simply stating whether the difference is statistically significant says little about the ecological importance of the difference.

which have been the subject of much recent attention (e.g., Downes et al. 2002). In the simplest BACI design, a single impact site and a single control (or reference) site are selected for study. The control site should closely match the impact site, except that the impact must not affect the control site. The purpose of the control site is to show what temporal changes would have occurred at the impact site had the impact not occurred. Control sites need not be on the same stream as the impact site, as long as you can reasonably expect their mussel communities to resemble those at the impact site, if the impact had not occurred. Both sites are studied once before the impact occurs and once after the impact occurs. For example, suppose we are concerned with the effect of the removal of a small dam on mussel populations downstream of the dam. We set up a study site directly below the dam and a similar study site (similar in terms of mussel populations, hydraulic and geomorphic variables) 3 km upstream of the dam. A week before the dam is removed, we estimate mussel densities in each study area (using one of the approaches described in "Determining Population Density or Size within a Defined Area"). The dam is removed, and we return 2 years later (long enough for the hypothesized impacts of dam removal to have occurred) and recensus the two sites. Computational details are shown in Box 11.

A simple BACI design like this has several shortcomings. In particular, the use of only a single control and impact site, and single measurement before and after the impact, makes a simple BACI design susceptible to idiosyncratic events unrelated to the impact being studied. For example, between the pre- and postimpact surveys, a tree may have fallen in the water at our single reference site, causing part of a mussel bed to be eroded away. Better designs use multiple control sites (and multiple impact sites, where possible) to avoid such idiosyncratic effects. Using multiple control sites provides a more robust estimate of what would have been expected to occur at the impact site in the absence of impacts. It may also be desirable to make multiple measurements before and after the impact, although this problem is less severe for long-lived mussels than for organisms with shorter life cycles. Going back to our study of the effects of dam removal, a better design would be to add several control sites, either far enough up- or downstream of the dam to be away from its impacts or on nearby streams. If several dams were being removed, it would be good to add impact sites below each of the dams. Downes et al. (2002) present a useful evaluation of several kinds of BACI designs. Again, it often will be useful to calculate and report the actual difference between control and impact sites over time (and the confidence limits on that difference) and not simply state that this difference is significant at some α-value.

Detecting Temporal Change in Mussel Populations

This is one of the most common objectives of mussel surveys but is surprisingly difficult to address satisfactorily. We will consider four approaches to detect whether mussel populations have changed: presence–absence, timed searches, quantitative

BOX 11. An Example of a BACI Analysis

Suppose we are interested in determining whether construction of a golf course along a stream changes population densities of mussels in the stream. A simple design for analyzing this problem is called the BACI design. The simplest such design (although not the best) would be to sample the mussel populations adjacent to the golf course and at a nearby control site before and after the golf course is built. Suppose the stream next to the golf course is 20 m wide, 0.4 m deep, gravel-bottomed, and supports eight species of mussels. We choose a control site (one that is as similar as possible to the impact site but not affected by the construction of the golf course) that supports seven species of mussels, looks a lot like the impact site, but is 1.5 km upstream of the golf course—we feel confident that the impacts of construction will not reach this site. On a fine June day, a month before construction begins, we sample both sites, using visual searches of ten 0.25-m^2 quadrats/site (we use far fewer samples than would be desirable in a real assessment to keep this example simple). Two years later, when we believe that the impacts of golf-course construction are complete, we repeat the survey. The data we collect are shown in the following table.

Impact, before	Impact, after	Control, before	Control, after
1	0	4	4
3	2	7	0
2	0	0	6
0	0	0	2
1	1	3	2
7	0	2	1
0	0	1	1
0	4	3	5
5	2	4	2
2	0	2	3

The mean density in the impact site fell from 8.4/m^2 to 3.6/m^2, whereas the mean density in the control site was unchanged (10.4/m^2), so it looks like there might have been an impact. To test for this impact, we transform the data as $\sqrt{(X + 0.5)}$ (see Box 10) and assess the significance of the time*site interaction term in a one-way ANOVA. The analysis follows.

Factor	F	p
Site	4.2	0.05
Time	0.9	0.35
Site × Time	1.2	0.28

Because the interaction term is insignificant at $\alpha = 0.05$, we conclude that there is only weak evidence for an impact from the golf course (an obvious result of using so few quadrats).

(Box continues on the following page.)

BOX 11 (continued)

Although widely used to assess impacts, the simple BACI design is not especially strong. Downes et al. (2002) discussed several improvements that can substantially strengthen the basic BACI design. For mussel populations, the most important of these probably is the use of multiple control sites (and multiple impact sites, if possible). Downes et al. (2002) should be consulted for the details of these designs and their analysis.

It appears that most BACIs have been based on simple random sampling within sites; but, there is no reason, in principle, why other sampling designs could not be used.

surveys (i.e., quadrats or transects) of plots that are not reused, and quantitative surveys of permanent plots.

Presence–absence data are often the only kind of information that is available about mussel populations, especially before 1990. Presence–absence data can be used to test for temporal change in mussel populations under certain conditions. (Technically, presence–absence data can test the hypothesis that the number of sites at which a species was detected has changed, not that the number of mussels in the population has changed, but the two are closely related in most cases). First, presence–absence data must be available from multiple sites; data from a single site are difficult to interpret as evidence of population change. Data from multiple sites are readily tested for temporal change, at least in principle. Consider n sites that were surveyed for the presence of a mussel species. At the time t_1, the species was found at $n_{1\bullet}$ sites and not found at $n_{2\bullet}$ sites (i.e., $n = n_{1\bullet} + n_{2\bullet}$). When the same n sites were resurveyed at a later time t_2, the species was found at n_{11} of the $n_{1\bullet}$ sites and n_{21} of the $n_{2\bullet}$ sites. The species was not found at n_{12} of the $n_{1\bullet}$ sites and n_{22} of the $n_{2\bullet}$ sites. The quantities n_{11} and n_{22} are uninformative as evidence of temporal change because species occurrence did not change at these sites. So, the question of temporal change is whether n_{21} (the number of sites where the species newly appeared) = n_{12} (the number of sites where the species disappeared). The data are tested with a χ^2 test, as follows:

$$\chi^2 = \frac{(|n_{21} - n_{12}| - 1)^2}{n_{21} + n_{12}}$$

(Snedecor and Cochran 1967: 214). An example is shown in Table 4.

Two important complications may arise in using presence–absence data to test for population change. Most obviously, the sampling methods in the two (or more) sampling times should be matched as closely as possible. If biologists spent three person-hours searching a site with mask-and-snorkel in August 1976, you should likewise use something like three person-hours of effort with mask-and-

TABLE 4. An example using presence–absence data to test for population decline. Sixty-seven sites in the Susquehanna River basin were surveyed for the green floater *Lasmigona subviridis* in 1955–1965 and again in 1996–1997 (from Strayer and Fetterman 1999). The hypothesis of no change in occurrence of this species cannot be rejected from this analysis.

$$\chi^2 = \frac{(|n_{21} - n_{12}| - 1)^2}{n_{21} + n_{12}} = \frac{(|7 - 6| - 1)^2}{7 + 6} = 0,\ p > 0.5.$$

1955–1965	1996–1997
Present ($n_{1\bullet}$) 12 sites	Still present (n_{11}) 6 sites
Absent ($n_{2\bullet}$) 55 sites	Disappeared (n_{12}) 6 sites
	Appeared (n_{21}) 7 sites
	Still absent (n_{22}) 48 sites

snorkel in August if you intend to test whether mussel populations have changed since 1976. A modification to this rule may be useful if your hypothesis is one-tailed (i.e., if you are postulating a direction of change). Thus, if you are testing whether populations of the wavyrayed lampmussel have declined (not just changed) in the Thames River basin, you may use methods that are as good or better than those used in the past. Such a procedure will provide a conservative test of decline. That is, if you detect a decline using this modified approach, you can be confident that the population actually declined. The disadvantage to using such a conservative approach is apparent if you do not detect a decline. In this case, you do not know whether the population is stable or if it really declined and your more intensive survey methods compensated for the decline. Thus, if you are interested in detecting temporal change, it generally is best to match sampling methods with the historical surveys that are the basis of comparison.

A second serious but less obvious problem concerns the selection of sampling sites. (This problem applies to all methods of assessing temporal change, not just presence–absence surveys, but may be especially severe for these surveys.) When presence–absence surveys are used to detect population change, usually the target population is something like the population of a mussel in a geographic area (e.g., a drainage basin or state). Nevertheless, resurveys of species status often are focused only on "historic sites"; that is, sites where the species was detected in a previous survey. If the resurvey is restricted to these historic sites, there are only two possible outcomes: that the species is rediscovered at all of the historic sites, in which case the population is said to be stable, or it is not rediscovered at all the historic sites, in which case the population is said to be declining. The problem here is that the sites resampled do not fairly represent the target population of interest (i.e., the historic sites do not represent the entire basin because they exclude sites where the species might occur but where the species was not found in the past). As a result, the conclusions of a survey restricted to historic sites correctly apply only to the historic sites, not to the whole geographic area of real interest. These can be quite different things, as shown by Table 4. In the upper Susquehanna basin, the green floater disappeared from half of its known historic sites between 1955–1965 and 1996–1997, leading to the erroneous conclusion (Strayer and Jirka 1997) that the species had declined in the basin. However, when the data were correctly analyzed over a broader range of sites in the basin, we found that the species did not decline at all, because it appeared at seven sites from which it was not known historically (Strayer and Fetterman 1999). The key here is that the resurvey must include sites where the species potentially occurs, not just sites where it has been found in the past.

Especially since about 1980, catch rates from timed searches have been widely used to assess mussel populations and to test whether mussel populations are changing over time. Despite their wide use, timed search data present special problems for assessing temporal change. The statistical analysis of timed search data is straightforward. If one site is sampled repeatedly, the data can be analyzed by regression (if a steady trend in abundance is expected), or a *t*-test or randomized intervention analysis (Carpenter et al. 1989) if a step-change in abundance is expected (and if temporal autocorrelation is not too severe). If many sites are visited twice, the data can be analyzed using a paired *t*-test, following data transformation if needed. Nevertheless, there are serious, unresolved issues with timed search data revolving around the underlying assumption that timed search catch rates are related to actual mussel population densities. To begin with, it appears that the relationship between mussel density and catch rate is very noisy (Figure 17), although this relationship has been explored only once (Strayer et al. 1997). The high variance around this relationship means that the timed search data may have a high variance and, thus, low power to detect temporal change.

Even more serious than this imprecision are several potential sources of large but uninvestigated bias. First, different observers sampling the same mussel population may report different catch rates because of their different search strategies and skills. This difference has not been investigated, but we believe that differences among experienced observers may cause as high as 3- to 10-fold variation in catch rates (Figure 20). Differences in catch rates among observers pose a special problem for programs to detect temporal change, which often use multiple observers who change over the course of the study. Second, the relationship between mussel density and catch rate may change over repeated visits to a site as a single observer becomes more familiar with the locations and habitat requirements of mussel beds. This problem ought to lead to catch rates that rise, then asymptote, as a single observer repeatedly visits a sampling site where a stable mussel population lives. This problem has not been quantified but should lead to severe bias. Third, changing conditions at a sampling site may alter the relationship between mussel density and catch rate. For instance, suppose a stable mussel population occupies a site that becomes increasingly turbid as a result of agricultural runoff. This increase in turbidity might be expected to lead to lower catch rates (Strayer et al. 1997; Smith et al. 2000) and to the erroneous conclusion that

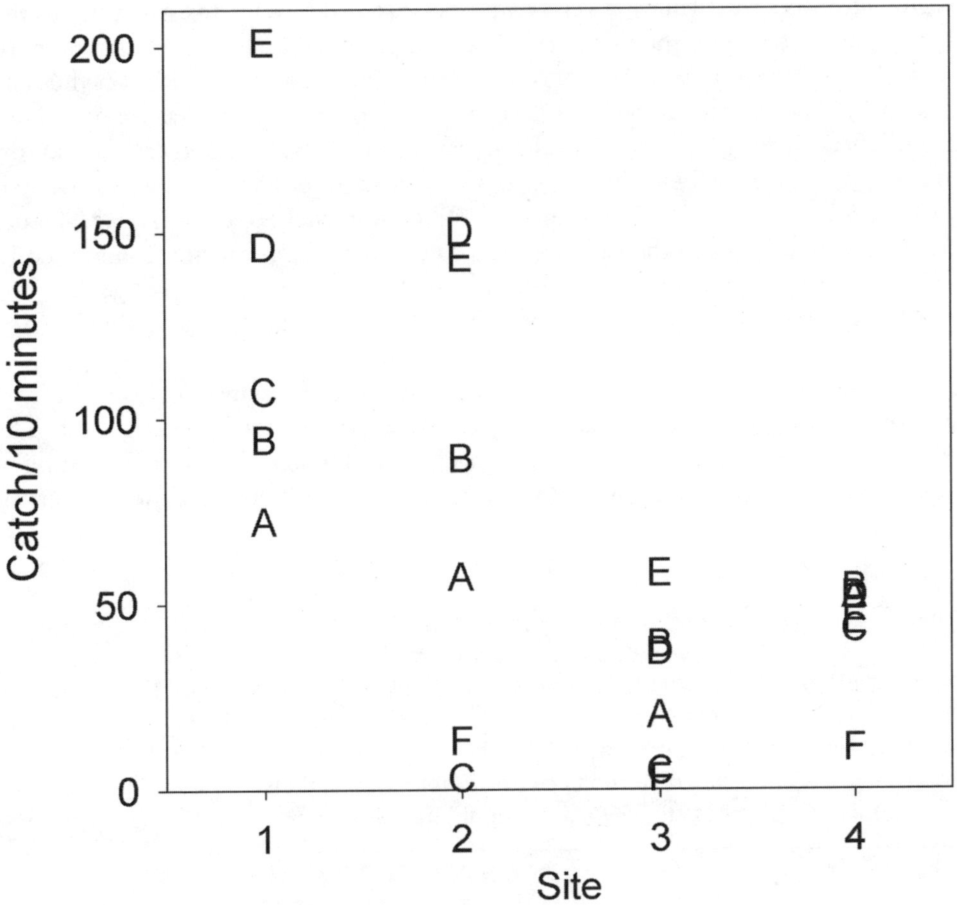

FIGURE 20. Differences in catch rates of mussels by several experienced observers (represented by the letters A–F) searching for mussels in the same reaches of stream at the same time. Note that catch rates vary greatly across observers and that the ordering of observers by catch rate is not consistent across sites. From Strayer and others (unpublished data).

agricultural runoff caused mussel populations to decline. These problems with timed search data are unlikely to be solved without excessive calibration studies, so it probably is best to avoid using timed searches to monitor temporal change in mussel populations.

It is possible to test for temporal change in mussel populations using quantitative data from quadrats or transects. Such data can be collected from permanent plots that are revisited over time or from two or more essentially independent surveys of an area. In the latter case, mussel populations are assessed at one time, using any appropriate method (see "Determining Population Density or Size within a Defined Area"), then reassessed at a later time, not necessarily using the same sampling plots. This will be a very common case if different investigators do the surveys, for example, if the first survey was not intended as a temporal baseline. In such a case, the simplest approach is to use a simple, unpaired t-test on the data (Table 5). The statistical power of this approach to detect temporal change is not high, especially for the sparse mussel populations that often are subject of monitoring programs (Figure 14). The ability of this approach to detect temporal change is weakened by the typically high spatial variability in mussel populations (e.g., Figure 3), which contributes to the variance used to calculate the t-statistic. Consequently, use of permanent plots may be a stronger way to detect temporal change.

Mussel populations may be monitored using permanent plots that can be resampled over time. The advantage of this approach is that some of the often-substantial spatial variation in mussel populations can be removed from the data before it is subjected to a statistical test, resulting in a more powerful test. The statistical test (a paired t-test) is run on the change in mussel densities at a sampling site, rather than in the densities themselves (Table 5). Note that the permanent

TABLE 5. Using quantitative data to detect temporal changes in a mussel population (eastern elliptio in the Hudson River estuary). To demonstrate the strength of permanent plots, which allow spatial variability to be reduced by sample pairing, we analyze these data with both unpaired and paired t-tests, following log-transformation of the data (see Box 10). The data are tested with a one-tailed t-test because the hypothesis is that the invasion of zebra mussels *Dreissena polymorpha* caused *Elliptio* populations to decline (not merely to change). From Strayer et al. (1994, and unpublished).

	Density (number/m^2)				
Site	1991–1992	1991–1992 ($\log_{10}[x + 1]$)	1997	1997 ($\log_{10}[x + 1]$)	Change in logged densities
A	57.4	1.77	9.6	1.03	–0.74
B	51.3	1.72	2.9	0.59	–1.13
C	38.3	1.59	30.6	1.50	–0.09
D	59.7	1.78	36.4	1.57	–0.21
E	0	0	0	0	0
F	0	0	0	0	0
G	2.3	0.52	2.9	0.59	0.07
H	2.3	0.52	0	0	–0.52
I	1.0	0.30	1.9	0.46	0.16
J	6.7	0.89	1.0	0.30	–0.59
K	0	0	0	0	0
L	0	0	0	0	0
			Statistics		
$\bar{x}$		0.757		0.503	–0.254
SE		0.219		0.168	0.115
t				–1.25 (unpaired)	–2.21 (paired)
p (1-tailed)				0.11	0.025

plots need not be precisely relocated to be useful; in the data shown in Table 5, we used loran to relocate points only ±60 m, but still were able to increase the sensitivity of the test for temporal change. Obviously, the more precisely the sample can be relocated, the more powerful the pairing technique.

The statistical power of permanent plot data to detect population change will depend on the variance of the change in mussel numbers, a property that seems not to have been investigated. The effectiveness of permanent plots will depend on the amount of patchiness in mussel communities, the precision with which plots can be relocated, and the amount of movement by mussels into and out of plots. We have a few data that bear on this problem from studies of two small upland streams and the Hudson River in New York. In the small upland streams (10–40 m wide), we marked the location of more than 200 1-m^2 plots on the stream bottom by driving stakes into both stream banks and stretching a fiberglass tape measure between the stakes. These plots, which could be relocated fairly precisely, were sampled twice using visual searches a year apart. The precision of estimates of population change derived from these permanent plots was ~10% greater than that derived from using unpaired plots and tended to increase with the density of mussel populations. In the Hudson River, we sampled mussel populations annually for nine years using a PONAR grab at 44 sites that were relocated using loran (with a precision of ±60 m). Here, the precision of estimates of population change from resampling these permanent sites was ~25% higher than would have been possible using unpaired plots and was unrelated to population density. The potential for permanent plots to contribute to assessment of temporal change in mussel populations needs further investigation.

Although it is most common to have only two time-points by which to assess temporal change, occasionally, an investigator will have multiple time-points (e.g., annual surveys for 10 years). In such cases, an investigator may use techniques such as regression analysis (Figure 21), randomized intervention analysis (Carpenter et al. 1989), or even time-series analysis (e.g., Box et al. 1994; Brockwell and Davis 2002) to assess temporal change and generally will have much greater statistical power to detect change (or estimate rates of change precisely) than an investigator limited to two time-points.

Finally, we note that the same concerns about simple significance testing we raised about analyses of mussel populations across sites apply equally to analyses of temporal changes in mussel populations. It often is more informative to calculate and present mean values, variances, and confidence intervals of estimates of temporal change than simply to determine if a temporal change is significant at an arbitrary α-value.

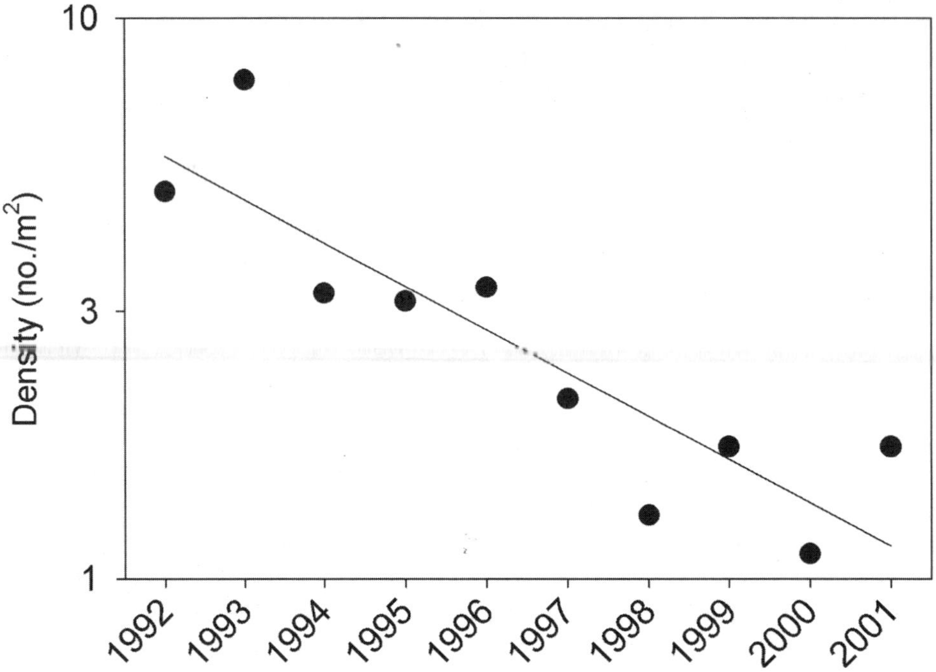

FIGURE 21. Time-course of loss of the eastern elliptio in the freshwater tidal Hudson River (study area = 140 km^2), showing an example of regression analysis of decline. Note the logarithmic scale on the *y*-axis; $r^2 = 0.78$, $p < 0.001$. The slope is –0.077, with 90% confidence limits of –0.127 to –0.027, corresponding to annual rates of decline of 16% (6–25%). From Strayer and Smith (1996, unpublished data).

Concluding Remarks

We have shown by now that no single one-size-fits-all sampling protocol will be suitable for all occasions. Instead, the best sampling protocol for your situation will depend on the five considerations with which we opened this guide and will vary widely from one situation to the next. We have tried to offer a wide range of sampling designs and methods from which to choose. Putting together a good sampling protocol is a creative process, which will require you to make deliberate decisions. Some of these decisions will be easy, while others may be very difficult. Nevertheless, we believe that it is worthwhile to work hard on your sampling protocol. Think about how much effort you will be spending first in the field and later interpreting your results, and how important it will be for your study to reach defensible conclusions. Good sampling designs and methods will make your money and effort go further and will make your inferences and conclusions more solid.

Further, the best available sampling designs will change over time, as research into mussel biology, collection methods, and sampling designs yields new information. We note, briefly, some of the important research areas that may affect your choice of sampling designs and methods in the future. Our knowledge of the burial behavior of mussels is recent (e.g., Amyot et al. 1991; Balfour and Smock 1995; Rogers et al. 2001) and incomplete. As we find out more about how different species, life stages, and sexes bury themselves in different seasons and different habitats, we will be better able to design sampling programs to include the entire mussel population. Our knowledge of the spatial structure of mussel populations (at scales from m to km) is very incomplete. Better information on the spatial structure of mussel populations (including variation across habitats and species) would allow us to design more effective sampling programs and would allow us to use simulation models to assess the efficiency of various sampling designs. Although we now know that careless handling can kill mussels that are being relocated (e.g., Cope and Waller 1995), we know essentially nothing about whether routine handling might affect mussels, especially populations of rare species that are frequently visited by researchers. If routine handling harms mussels, then we will need to rethink some of our monitoring and research activities. Most sampling methods in current use (visual searches, hand-excavation) are not well suited for large, deep lakes and rivers. We could benefit from research on the utility of remote methods (sampling by grabs, dredges, and hydroacoustic methods) in sampling mussel populations in large lakes and rivers. Although adaptive designs were invented specifically for patchy and rare organisms, and would seem to be well suited to mussels, initial tests of adaptive methods on mussel populations have been disappointing (Strayer et al. 1997; Smith et al. 2003). It would be useful to have more evaluations of adaptive designs under field conditions. In general, we could use more information on the costs and precision of different sampling designs under field conditions so that we could more objectively compare the value of different designs in different settings. It would help if biologists more often reported the costs of their sampling programs, at least in terms of labor. If biologists are going to continue to use timed searches and other informal methods, it is essential that these methods be subjected to rigorous tests, so that we can better understand the conditions under which these methods can provide useful information. Initial tests of these methods (Figure 17) have not been promising. Finally, ecologists are actively developing and testing variants on the basic BACI design to detect impacts (e.g., Downes et al. 2002). Mussel biologists interested in assessing impacts will need to stay abreast of this rapidly developing literature. As a result of progress in these and other areas, recommendations for sampling designs will evolve over time.

Appendix: Learning to Identify Freshwater Mussels

Most surveys of mussel populations demand accurate identification of the mussels. It is beyond the scope of this guide to teach mussel identification, but we will offer general advice about how to learn to identify freshwater mussels and provide an entry into the literature on mussel identification.

Unionid mussels are not particularly difficult to identify, but they can be difficult and frustrating for a novice. Compared to freshwater plants, insects, and fish, many characters used to identify mussels are vague and subjective rather than clear and quantifiable. For instance, keys for identifying fish and insects often rely on characters that are unambiguously present or absent (e.g., various thoracic sclerites in caddisfly larvae) or countable (e.g., the number of fin rays in fish). In contrast, mussel keys are filled with subjective characters like "shell squarish" versus "shell roundish." The only way to learn the difference between "roundish" and "squarish" is to examine specimens (or good illustrations) of the pertinent species. Further, the shells of mussels (upon which most routine identification is based) can be dismayingly variable: different individuals of the same species may differ greatly in size, shape, and color. Thus, it is difficult to learn mussel identification from a book, and beginners are likely to misidentify many mussels.

To learn mussel identification, it is essential to have practical experience handling many specimens of each species, preferably in the company of someone who already knows mussel identification and can offer tips for identification. A good place to start is a short course with an expert. Many museums and field stations now offer short courses in mussel identification. Alternatively, most mussel biologists are happy to share their knowledge, and it may be possible to arrange to spend a few days in the laboratory or field with someone who is familiar with the local fauna. Finally, it sometimes is possible to gain access to a curated collection of mussels at a museum. By examining a large series of shells with your favorite guidebook in your hand, you can begin to learn the essential characters and appearance of each species. Ultimately, you will learn simply to recognize the species by sight, without using a key.

Despite our admonition that mussel identification is difficult to learn solely from a book, a good identification guide can be an enormous help and may offer useful information about techniques and mussel biology. Generally, the smaller the geographic scope of the guide, the easier it will be to use. Thus, if you are trying to learn the mussels of Lake Champlain, it will be easier to use a guide to the mussels of Vermont than one that covers all of the United States or North America. (The key of McMahon and Bogan, although continental in scope, is divided into four geographic regions, making it easier to use than a single continental key.)

The literature on the identification, taxonomy, systematics, and nomenclature of freshwater mussels is very large (see *http://ellipse.inhs.uiuc.edu:591/mollusk/biblio.html* for a searchable bibliography of the worldwide literature on freshwater mussels). Below, we provide a list of a few useful references for the identification of freshwater mussels in North America. In addition, much useful material is posted on the Web. Rather than providing what will inevitably be an outdated list of Internet resources, we suggest that you follow the links from the more permanent Web sites of leading scientific societies (in North America: the Freshwater Mollusk Conservation Society, *http://ellipse.inhs.uiuc.edu/FMCS/;* the American Malacological Society, *http://data.acnatsci.org/ams;* the Conchologists of America, *http://coa.acnatsci.org/conchnet;* and the North American Benthological Society, *http://www.benthos.org*).

Baker, F. C. 1928. The freshwater Mollusca of Wisconsin, part II: Pelecypoda. Bulletin of the Wisconsin Geological and Natural History Survey 70:1–495.

Bequaert, J. C., and W. B. Miller. 1973. The mollusks of the arid Southwest with an Arizona check list. University of Arizona Press, Tucson.

Bleam, D. E., K. J. Couch, and D. A. Distiller. 1999. Key to the unionid mussels of Kansas. Transactions of the Kansas Academy of Science 102: 83–91.

Brandauer, N., and S. K. Wu. 1978. The Bivalvia of Colorado. University of Colorado Museum, Boulder.

Brim Box, J., and J. D. Williams. 2000. Unionid mollusks of the Apalachicola Basin in Alabama, Florida, and Georgia. Bulletin of the Alabama Museum of Natural History 21:1–143.

Burch, J. B. 1975. Freshwater unionacean clams (Mollusca: Pelecypoda) of North America, revised edition. Malacological Publications, Hamburg, Michigan.

Burch, J. B., and C. M. Patterson. 1976. Key to the genera of freshwater pelecypods (mussels and clams) of Michigan. University of Michigan Museum of Zoology Circular 4.

Chamberlin, R. V., and D. T. Jones. 1929. A descriptive catalog of the Mollusca of Utah. University of Utah Bulletin 19:1–203.

Clarke, A. H. 1981a. The freshwater molluscs of Canada. National Museums of Canada, Ottawa.

Clarke, A. H. 1981b. The tribe Alasmidontini (Unionidae: Anodontinae), part I: *Pegias*, *Alasmidonta*, and *Arcidens*. Smithsonian Contributions to Zoology 326:1–101.

Clarke, A. H. 1985. The tribe Alasmidontini (Unionidae: Anodontinae), part II: *Lasmigona* and *Simpsonaias*. Smithsonian Contributions to Zoology 399:1–75.

Cordeiro, J. R. 1999. Distribution and habitat of freshwater mussels (Bivalvia: Unionoida: Unionidae) in Colorado. University of Colorado Museum, Natural History Inventory of Colorado 19, Boulder.

Couch, K. J. 1997. An illustrated guide to the unionid mussels of Kansas. Karen J. Couch, Olathe, Kansas.

Cummings, K. S., and C. A. Mayer. 1992. Field guide to freshwater mussels of the Midwest. Illinois Natural History Survey Manual 5:1–194.

Cvancara, A. M. 1983. Aquatic mollusks of North Dakota. North Dakota Geological Survey Report of Investigation 78:1–141.

Fichtel, C., and D. G. Smith. 1995. The freshwater mussels of Vermont. Vermont Fish and Wildlife Department Technical Report 18:1–54.

Goodrich, C. 1932. The Mollusca of Michigan. University of Michigan Press, Michigan Handbook Series 5, Ann Arbor.

Harris, J. L., and M. E. Gordon. 1990. Arkansas mussels. Arkansas Game and Fish Commission, Little Rock.

Heard, W. H. 1979. Identification manual of the freshwater clams of Florida. Florida Department of Environmental Regulation Technical Series 4:1–83.

Howells, R. G., R. W. Neck, and H. D. Murray. 1996. Freshwater mussels of Texas. Texas Parks and Wildlife Press, Austin.

Ingram, W. M. 1948. The larger freshwater clams of California, Oregon, and Washington. Journal of Entomology and Zoology 40:72–92.

Johnson, R. I. 1972. The Unionidae (Mollusca: Bivalvia) of peninsular Florida. Bulletin of the Florida State Museum, Biological Sciences 16:181–249.

Johnson, R. I. 1977. Monograph of the genus *Medionidus* (Bivalvia: Unionidae) mostly from the Apalachicolan region, southeastern United States. Occasional Papers on Mollusks, Museum of Comparative Zoology 4:161–187.

Johnson, R. I. 1978. Systematic and zoogeography of *Plagiola* (=*Dysnomia* =*Epioblasma*), an almost extinct genus of freshwater mussels (Bivalvia: Unionidae) from middle North America. Bulletin of the Museum of Comparative Zoology 148:239–321.

Mathiak, H. A. 1979. A river survey of the unionid mussels of Wisconsin 1973–1977. Sand Shell Press, Horicon, Wisconsin.

McMahon, R. F., and A. E. Bogan. 2001. Mollusca: Bivalvia. Pages 331–429 *in* J. H. Thorp and A. P. Covich, editors. Ecology and classification of North American freshwater invertebrates, 2nd edition. Academic Press, San Diego, California.

Murray, H. D., and A. B. Leonard. 1962. Handbook of unionid mussels in Kansas. Miscellaneous Publications of the Museum of Natural History, University of Kansas 28:1–184.

Nedeau, E. J., M. A. McCollough, and B. I. Schwartz. 2000. The freshwater mussels of Maine. Maine Department of Inland Fisheries and Wildlife, Augusta.

Oesch, R. D. 1995. Missouri naiades: a guide to the mussels of Missouri, 2nd edition. Missouri Department of Conservation, Jefferson City.

Parmalee, P. W. 1967. The freshwater mussels of Illinois. Illinois State Museum Popular Science Series 8:1–108.

Parmalee, P. W., and A. E. Bogan. 1998. The freshwater mussels of Tennessee. University of Tennessee Press, Knoxville.

Smith, D. G. 1995. Keys to the freshwater macroinvertebrates of Massachusetts. D. G. Smith, Sunderland, Massachusetts.

Smith, D. G. 2001. Pennak's freshwater invertebrates of the United States, 4th edition. Wiley, New York.

Stern, E. M. 1990. An illustrated key to the freshwater mussels (Bivalvia: Unionidae) of Wisconsin. University of Wisconsin at Stevens Point Reports of the Museum of Natural History 20.

Strayer, D. 1990. Freshwater Mollusca. Pages 335–371 *in* B. L. Peckarsky, P. R. Fraissinet, M. A. Penton, and D. J. Conklin, editors. Freshwater macroinvertebrates of northeastern North America. Cornell University Press, Ithaca, New York.

Strayer, D. L., and K. J. Jirka. 1997. The pearly mussels of New York State. Memoirs of the New York State Museum 26: 1–113.

Turgeon, D. D., J. F. Quinn, A. E. Bogan, E. V. Coan, F. G. Hochberg, W. G. Lyons, P. M. Mikkelsen, R. J. Neves, C. F. E. Roper, G. Rosenberg, B. Roth, A. Scheltema, F. G. Thompson, M. Vecchione, and J. D. Williams. 1998. Common and scientific names of aquatic invertebrates from the United States and Canada: mollusks, 2nd edition. American Fisheries Society, Special Publication 26, Bethesda, Maryland.

Valentine, B. D., and D. H. Stansbery. 1971. An introduction to the naiades of the Lake Texoma region, Oklahoma, with notes on the Red River fauna (Mollusca: Unionidae). Sterkiana 42:1–40.

Vidrine, M. F. 1993. The historical distributions of freshwater mussels in Louisiana. Gail Q. Vidrine Collectibles, Eunice, Louisiana.

Watters, G. T. 1993. A guide to the freshwater mussels of Ohio, revised edition. Ohio Division of Wildlife, Columbus.

Glossary

Accuracy.—the difference between the estimated value of a parameter and its true value. If the true and estimated values of a parameter are close, then accuracy is high; if they are far apart, accuracy is low. Accuracy depends on bias and precision (see definitions of *bias* and *precision*).

Bias.—the difference between the true population parameter and its expected estimated value. If this difference is large, the estimate is biased. For example, estimates of mussel population size may be biased low if buried animals are not sampled effectively.

Confidence interval.—an estimated range that represents plausible values for a population parameter. Thus, a 95% confidence interval (if correctly estimated) includes the true value of the parameter 95% of the time.

Cost.—the amount of money, labor, and other resources needed to complete a specified job. In calculations of optimal sampling designs, cost usually is expressed as the amount of money or labor expected to be needed for a task.

CPUE (catch per unit effort).—the number of mussels encountered per unit effort expended in sampling, usually expressed as the number of mussels per person-hour of searching. The results of *timed searches* are usually given as CPUE.

Density.—the number of mussels per unit area, usually given as number/m^2.

Design.—the procedure by which an investigator chooses which units of the population to sample.

Estimation.—the process whereby sample data are used to estimate a population parameter.

False change (type I error).—the probability of incorrectly rejecting the hypothesis of no population change (also called a false positive), as determined by the α-value. If $\alpha = 0.05$, you have a 5% chance of making a type I error.

Inference.—drawing general conclusions from specific results. Statistical inference is drawing conclusions about a population based on data from a sample.

Mean.—a measure of the average value of a variable. We use "mean" to refer to the arithmetic mean, $\Sigma X_i/n$, where X_i are the individual observations (e.g., number of mussels) made on each of n sampling units.

Missed change (type II error).—the probability of failing to detect a true population change, also called a false negative. Type II error rate depends on the magnitude of the change, the underlying variability in the change, the type I error rate, and the sample size. Sample size in the context of monitoring a population includes both how many time periods that a population is monitored and how much sampling effort is expended with each time period. The complement of type II error rate is statistical power (i.e., Power = 1 – type II error).

Parameter.—a quantity representing an attribute of a population. Examples are population mean, variance, density, and total.

Population.—a collection of discrete objects or observations with well defined boundaries in time and space, about which you are trying to make inferences. Examples of statistical populations include all nonoverlapping 1-m^2 sections of stream bottom between two bridges in the summer of 1997, or the lengths of all male *Epioblasma triquetra* in that same area. Note that statisticians use "population" in a different sense than ecologists, who mean all individuals of a species in a defined area, or geneticists, who mean all individuals in a defined area who exchange genetic information.

Power.—see *Statistical power.*

Precision.—the repeatability of an estimate from one sample to the next. Precision usually is measured by sampling variance or by standard error. Estimates may be very precise but still be inaccurate because of *bias* (e.g., Box 8)

Primary sampling unit.—in two-stage sampling, the sampling units that are selected at the first stage of sampling.

Quadrat.—a rectangular frame used to define the boundaries of an area to be sampled (Figure 11).

Random sample.—a sample in which each unit in the population has an equal chance of being chosen for sampling. Random sampling requires a formal design and is not the same as haphazard sampling, in which areas to be sampled are chosen without a formal plan.

Sample.—a collection of elements or units representing a subset of all units in a population.

Sampling distribution.—the values that a statistic will take in all possible samples from a population.

Sampling error.—a term used loosely to describe the variability of an estimate from one sample to the next. Sometimes used synonymously with standard error.

Sampling unit.—an element or unit of a population defined spatially and temporally that can be selected independently of all other units.

Secondary sampling unit.—in two-stage sampling designs, the sampling units that are selected at the second stage of sampling.

Standard deviation.—the square root of variance. Population standard deviation measures the variability among all units in a population. Sample standard deviation measures variability among all units in a sample.

Standard error.—a measure of the reliability of an estimate—how variable an estimate is from one sample to the next. Standard error is the standard deviation of the sampling distribution. By definition, it is the population standard deviation divided by the square root of sample size, and it can be estimated as the sample standard deviation divided by the square root of sample size.

Statistic.—a quantity that can be calculated from sample data alone. Often a statistic is used to estimate a population parameter. For example, the sample mean is used to estimate the population mean.

Statistical power.—the probability of correctly rejecting a false null hypothesis. For example, in the context of monitoring a population, statistical power is the probability of detecting a true population change. Statistical power depends on the underlying variability in the change, the magnitude of the change, type I error rate, and sample size. Sample size in the context of monitoring a population includes both how many time periods that a population is monitored and how much sampling effort is expended with each time period. The complement of statistical power is the type II error rate (i.e., type II error = 1 – Power).

Survey.—a procedure used in an observational study to sample a population for the purpose of estimating a population parameter.

Timed search.—a procedure, usually informal, in which an investigator records the number of mussels found during a search of a known length of time, usually expressed as *catch-per-unit-effort* (*CPUE*).

Transect.—a long, narrow strip used to define the boundaries of the area to be sampled.

Type I error.—see *False change.*

Type II error.—see *Missed change.*

Uncertainty.—a term used synonymously with variation to represent the unpredictable variability of data and of any quantity that is computed from data.

Variable.—a measurable or observable attribute or characteristic of a population element or unit. For example, the count of mussels within a 0.25-m^2 quadrat is a variable.

Variance.—a measure of variability among elements or units. Population variance measures variability among all units in a population. Sample variance represents variability among units in a sample. Sampling variance represents variability of a statistic or estimate among all possible samples from a population.

Literature Cited

Amyot, J.-P., and J. A. Downing. 1991. Endo- and epibenthic distribution of the unionid mollusc *Elliptio complanata*. Journal of the North American Benthological Society 10:280–285.

Anderson, D. R., K. P. Burnham, and W. L. Thompson. 2000. Null hypothesis testing: problems, prevalence, and an alternative. Journal of Wildlife Management 64:912–923.

Balfour, D. L., and L. A. Smock. 1995. Distribution, age structure, and movements of the freshwater mussel *Elliptio complanata* (Mollusca: Unionidae) in a headwater stream. Journal of Freshwater Ecology 10:255–268.

Bayley, P. B., and J. T. Peterson. 2001. Species presence for zero observations: an approach and an application to estimate probability of occurrence of fish species and species richness. Transactions of the American Fisheries Society 130:620–633.

Berry, D. A. 1987. Logarithmic transformations in ANOVA. Biometrics 43:439–456.

Bogan, A. E. 1993. Freshwater bivalve extinctions (Mollusca: Unionoida): a search for causes. American Zoologist 33:599–609.

Bowen, Z. H., S. P. Malvestuto, W. D. Davies, and J. H. Crance. 1994. Evaluation of the mussel fishery in Wheeler Reservoir, Tennessee River. Journal of Freshwater Ecology 9:313–319.

Box, G. E. P., G. M. Jenkins, and G. C. Reinsel. 1994. Time-series analysis: forecasting and control, 3rd edition. Prentice Hall, Upper Saddle River, New Jersey.

Brockwell, P. J., and R. A. Davis. 2002. Introduction to time series and forecasting, 2nd edition. Springer-Verlag, New York.

Brownie, C., J. E. Hines, J. D. Nichols, K. H. Pollock, and J. B. Hestbeck. 1993. Capture–recapture studies for multiple strata including non-Markovian transition probabilities. Biometrics 49:1173–1187.

Buckland, S. T., D. R. Anderson, K. P. Burnham, and J. L. Laake. 1993. Distance sampling: estimating abundance of biological populations. Chapman and Hall, London.

Buckland, S. T., D. R. Anderson, K. P. Burnham, J. L. Laake, D. L. Borchers, and L. Thomas. 2001. Introduction to distance sampling: estimating abundance of biological populations. Oxford University Press, Oxford, UK.

Burnham, K. P., and D. R. Anderson. 2002. Model selection and multi-model inference: a practical information–theoretic approach, 2nd edition. Springer-Verlag, New York.

Burnham, K. P., D. R. Anderson, G. C. White, C. Brownie, and K. H. Pollock. 1987. Design and analysis methods for fish survival experiments based on release–recapture. American Fisheries Society, Monograph 5, Bethesda, Maryland.

Carlander, H. B. 1954. Mussel fishing and the pearl button industry. Pages 40–51 *in* A history of fish and fishing in the Upper Mississippi River. Upper Mississippi River Conservation Committee, Rock Island, Illinois.

Carpenter, S. R., T. M. Frost, D. Heisey, and T. K. Kratz. 1989. Randomized intervention analysis and the interpretation of whole-ecosystem experiments. Ecology 70:1142–1152.

Chao, A. 2001. An overview of closed capture-recapture models. Journal of Agricultural, Biological, and Environmental Statistics 6:158–175.

Christman, M. C. 2000. A review of quadrat-based sampling of rare, geographically clustered populations. Journal of Agricultural, Biological, and Environmental Statistics 5:168–201.

Christman, M. C., and J. S. Pontius. 2000. Bootstrap confidence intervals for adaptive cluster sampling. Biometrics 56:503–510.

Claassen, C. 1994. Washboards, pigtoes, and muckets: historic musseling in the Mississippi watershed. Historical Archaeology 28:1–145.

Cochran, W. G. 1977. Sampling techniques, 3rd edition. Wiley, New York.

Cohen, J. 1988. Statistical power analysis for the behavioral sciences, 2nd edition. Lawrence Erdman Associates, Hillsdale, New Jersey.

Cope, W. G., and D. L. Waller. 1995. Evaluation of freshwater mussel relocation as a conservation and management strategy. Regulated Rivers: Research and Management 11:147–155.

Cvancara, A. M. 1972. Lake mussel distribution as determined with SCUBA. Ecology 53:154–157.

Dorazio, R. M. 1999. Design-based and model-based inference in surveys of freshwater mollusks. Journal of the North American Benthological Society 18:118–131.

Downes, B. J., L. A. Barmuta, P. G. Fairweather, D. P. Faith, M. J. Keough, P. S. Lake, B. D. Mapstone, and G. P. Quinn. 2002. Assessing ecological impacts: concepts and practice in flowing waters. Cambridge University Press, Cambridge, UK.

Downing, J. A. 1979. Aggregation, transformation, and the design of benthos sampling programs. Journal of the Fisheries Research Board of Canada 36:1454–1463.

Downing, J. A. 1984. Sampling the benthos of standing waters. Pages 87–130 *in* J. A. Downing and F. H. Rigler, editors. A manual on methods for the assessment of secondary productivity in fresh waters, 2nd edition. Blackwell Scientific Publications, Oxford, UK.

Downing, J. A., J.-P. Amyot, M. Pérusse, and Y. Rochon. 1989. Visceral sex, hermaphroditism, and protandry in a population of the freshwater bivalve *Elliptio complanata*. Journal of the North American Benthological Society 8:92–99.

Downing, J. A., and W. L. Downing. 1992. Spatial aggregation, precision, and power in surveys of freshwater mussel populations. Canadian Journal of Fisheries and Aquatic Sciences 49:985–991.

Downing, J. A., and F. H. Rigler, editors. 1984. A manual on methods for the assessment of secondary productivity in fresh waters, 2nd edition. Blackwell Scientific Publications, Oxford, UK.

Dunn, H. L., B. E. Sietman, and D. E. Kelner. 2000. Evaluation of recent unionid (Bivalvia) relocations and suggestions for future relocations and reintroductions. Pages 169–183 *in* P. D. Johnson and R. S. Butler, editors. Proceedings of the first Freshwater Mollusk Conservation Society symposium. Ohio Biological Survey, Columbus.

Eberhardt, L. L., and M. A. Simmons. 1987. Calibrating population indices by double sampling. Journal of Wildlife Management 51:665–675.

Elder, J. F., and J. J. Collins. 1991. Fresh-water mollusks as indicators of bioavailability and toxicity of metals in surface-water systems. Reviews of Environmental Contamination and Toxicology 122:37–79.

Elliott, J. M. 1977. Some methods for the statistical analysis of samples of benthic invertebrates. Freshwater Biological Association Scientific Publication 25.

Elliott, J. M., and C. M. Drake. 1981. A comparative study of seven grabs used for sampling benthic macroinvertebrates in rivers. Freshwater Biology 11: 99–120.

Green, R. H., S. G. Hinch, J. L. Metcalfe, and V. H. Young. 1989. Use of freshwater mussels (Bivalvia, Unionidae) to monitor the nearshore environment of lakes. Journal of Great Lakes Research 15:635–644.

Green, R. H., and R. C. Young. 1993. Sampling to detect rare species. Ecological Applications 3:351–356.

Hanson, J. M., W. C. Mackay, and E. E. Prepas. 1988a. The effects of water depth and density on the growth of a unionid clam. Freshwater Biology 19:345–355.

Hanson, J. M., W. C. Mackay, and E. E. Prepas. 1988b. Population size, growth, and production of a unionid clam, *Anodonta grandis simpsoniana*, in a small, deep Boreal Forest lake in central Alberta. Canadian Journal of Zoology 66:247–253.

Hart, R. A., J. W. Grier, and A. C. Miller. 2001. Empirically derived survival rates of a native mussel, *Amblema plicata*, in the Mississippi and Otter Tail rivers, Minnesota. American Midland Naturalist 146:254–263.

Hedayat, A. S., and B. K. Sinha. 1991. Design and inference in finite population sampling. Wiley, New York.

Hoggarth, M. A., D. L. Rice, and T. L. Grove. 2000. The correlation of mussels with fish in the upper Blanchard River in Hardin and Hancock counties, Ohio, with special regard to the rayed bean (*Villosa fabalis*). Pages 19–26 *in* R. A. Tankersley, D. I. Warmolts, G. T. Watters, and B. J. Armitage, editors. Proceedings of the conservation, captive care, and propagation of freshwater mussels symposium. Ohio Biological Survey, Columbus.

Hornbach, D. J., and T. Deneka. 1996. A comparison of a qualitative and a quantitative collection method for examining freshwater mussel assemblages. Journal of the North American Benthological Society 15:587–596.

Huehner, M. K., and C. L. Corr. 1994. The unionid mussel fauna of Pymatuning Creek in Ashtabula County, Ohio, including the federally endangered clubshell (*Pleurobema clava*). Ohio Journal of Science 94:147–150.

Isom, B. G., and C. Gooch. 1986. Rationale and sampling designs for freshwater mussels Unionidae in streams, large rivers, impoundments, and lakes. Pages 46–59 *in* B. G. Isom, editor. Rationale for sampling and interpretation of ecological data in the assessment of fresh-water ecosystems. American Society for Testing and Materials, STP 894, Philadelphia.

Johnson, D. H. 1999. The insignificance of statistical significance testing. Journal of Wildlife Management 63:763–772.

Kendall, W. L., and R. Bjorkland. 2001. Using open robust design models to estimate temporary emigration from capture–recapture data. Biometrics 57:1113–1122.

Kendall, W. L., J. D. Nichols, and J. E. Hines. 1997. Estimating temporary emigration and breeding proportions using capture–recapture data with Pollock's robust design. Ecology 78:563–578.

Kendall, W. L., K. H. Pollock, and C. Brownie. 1995. A likelihood-based approach to capture–recapture estimation of demographics parameters under the robust design. Biometrics 51:293–308.

Krebs, C. J. 1998. Ecological methodology, 2nd edition. Addison-Wesley, Boston.

Lefevre, G., and W. C. Curtis. 1910. Reproduction and parasitism in the Unionidae. Journal of Experimental Zoology 9:79–116.

Lellis, W. A. 2001. Freshwater mussel survey of the Upper Delaware Scenic and Recreational River: qualitative survey 2000. Report to the National Park Service, Milanville, Pennsylvania.

MacKenzie, D. I., J. D. Nichols, G. B. Lachman, S. Droege, J. A. Royle, and C. A. Langtimm. 2002. Estimating site occupancy rates when detection probabilities are less than one. Ecology 83:2248–2255.

Master, L. L., B. A. Stein, L. S. Kutner, and G. A. Hammerson. 2000. Vanishing assets: conservation status of U.S. species. Pages 93–118 *in* B. A. Stein, L. S. Kutner, and J. S. Adams, editors. Precious heritage: the status of biodiversity in the United States. Oxford University Press, New York.

McArdle, B. H. 1990. When are species not there? Oikos 57:276–277.

McArdle, B. H., K. J. Gaston, and J. H. Lawton. 1990. Variation in the size of animal populations: patterns, problems, and artifacts. Journal of Animal Ecology 59:439–454.

Metcalfe, J. L., and M. N. Charlton. 1990. Fresh-water mussels as biomonitors for organic industrial contaminants and pesticides in the St. Lawrence River. Science of the Total Environment 97/98:595–615.

Metcalfe-Smith, J. L., J. Di Maio, S. K. Staton, and G. L. Mackie. 2000. Effect of sampling effort on the efficiency of the timed search method for sampling freshwater mussel communities. Journal of the North American Benthological Society 19:725–732.

Miller, A. C., and B. S. Payne. 1993. Qualitative versus quantitative sampling to evaluate population and community characteristics at a large-river mussel bed. American Midland Naturalist 130:133–145.

Miller, A. C., B. S. Payne, D. J. Shafer, and L. T. Neill. 1993. Techniques for monitoring freshwater bivalve communities and populations in large rivers. Pages 147–158 *in* K. S. Cummings, A. C. Buchanan, and L. M. Koch, editors. Conservation and management of freshwater mussels. Upper Mississippi River Conservation Committee, Rock Island, Illinois.

Miller, A. C., R. Whiting, and D. B. Wilcox. 1989. An evaluation of a skimmer dredge for collecting freshwater mussels. Journal of Freshwater Ecology 5:151–154.

Naimo, T. J. 1995. A review of the effects of heavy metals on fresh-water mussels. Ecotoxicology 4:341–362.

Nalepa, T. F., W. S. Gardner, and J. M. Malczyk. 1991. Phosphorus cycling by mussels (Bivalvia) in Lake St. Clair. Hydrobiologia 219:239–250.

Negus, C. L. 1966. A quantitative study of growth and production of unionid mussels in the River Thames at Reading. Journal of Animal Ecology 35:513–532.

Nelson, D. 1982. Sampling for mussels. Pages 41–60 *in* A. C. Miller, compiler. Report of freshwater mollusks workshop, 1981. U.S. Army Engineer Waterways Experiment Station, Vicksburg, Mississippi.

Neves, R. J., and J. C. Widlak. 1987. Habitat ecology of juvenile freshwater mussels (Bivalvia:Unionidae) in a headwater stream in Virginia. American Malacological Bulletin 5:1–7.

Nicholson, M., and J. Barry. 1995. Inferences from spatial surveys about the presence of an unobserved species. Oikos 72:74–78.

Obermeyer, B. K. 1998. A comparison of quadrats versus timed snorkel searches for assessing freshwater mussels. American Midland Naturalist 139:331–339.

Ortmann, A. E. 1911. A monograph of the najades of Pennsylvania. Memoirs of the Carnegie Museum 4:279–347.

Osenberg, C. W., R. J. Schmitt, S. J. Holbrook, K. E. Abu-Saba, and A. R. Flegal. 1994. Detection of environmental impacts: natural variability, effect size, and power analysis. Ecological Applications 4:16–30.

Otis, D. L., K. P. Burnham, G. C. White, and D. R. Anderson. 1978. Statistical inference for capture data on closed animal populations. Wildlife Monographs 62.

Payne, B. S., A. C. Miller, and R. Whiting. 1997. Designing a riverine mussel survey. Pages 150–156 *in* K. S. Cummings, A. C. Buchanan, C. A. Mayer, and

T. J. Naimo, editors. Conservation and management of freshwater mussels II. Initiatives for the future. Upper Mississippi River Conservation Committee, Rock Island, Illinois.

Peterman, R. M. 1990. Statistical power analysis can improve fisheries research and management. Canadian Journal of Fisheries and Aquatic Sciences 47:2–15.

Pollock, K. H., J. D. Nichols, J. E. Hines, and C. Brownie. 1990. Statistical inference for capture–recapture experiments. Wildlife Monographs 107.

Richardson, T. D., and P. Yokely. 1996. A note on sampling technique and evidence of recruitment in freshwater mussels (Unionidae). Archiv für Hydrobiologie 137:135–140.

Rogers, S. O., B. T. Watson, and R. J. Neves. 2001. Life history and population biology of the endangered tan riffleshell (*Epioblasma florentina walkeri*) (Bivalvia: Unionidae). Journal of the North American Benthological Society 20:582–594.

Salehi, M. M., and G. A. F. Seber. 1997. Two-stage adaptive cluster sampling. Biometrics 53:959–970.

Schwarz, C. J., and G. A. F. Seber. 1999. Estimating animal abundance: review III. Statistical Science 14:427–456.

Seber, G. A. F. 1982. Estimation of animal abundance, 2nd edition. Griffin, London.

Seber, G. A. F. 2001. Some new directions in estimating animal population parameters. Journal of Agricultural, Biological, and Environmental Statistics 6:140–151.

Skalski, J. R., and D. S. Robson. 1992. Techniques for wildlife investigations: design and analysis of capture data. Academic Press, New York.

Smith, D. R., R. F. Villella, and D. P. Lemarié. 2001. Survey protocol for assessment of endangered freshwater mussels in the Allegheny River, Pennsylvania. Journal of the North American Benthological Society 20:118–132.

Smith, D. R., R. F. Villella, and D. P. Lemarié. 2003. Application of adaptive cluster sampling to low-density populations of freshwater mussels. Environmental and Ecological Statistics 10:7–15.

Smith, D. R., R. F. Villella, D. P. Lemarié, and S. von Oettingen. 2000. How much excavation is needed to monitor freshwater mussels? Pages 203–218 *in* P. D. Johnson and R. S. Butler, editors. Proceedings of the first Freshwater Mollusk Conservation Society symposium. Ohio Biological Survey, Columbus.

Snedecor, G. W., and W. G. Cochran. 1967. Statistical methods, 6th edition. Iowa State University Press, Ames.

Starrett, W. C. 1971. A survey of the mussels (Unionacea) of the Illinois River: a polluted stream. Bulletin of the Illinois Natural History Survey 30:265–403.

Steidl, R. J., J. P. Hayes, and E. Schauber. 1997. Statistical power analysis in wildlife research. Journal of Wildlife Management 61:270–279.

Strayer, D. 1983. The effects of surface geology and stream size on freshwater mussel distribution in southeastern Michigan, USA. Freshwater Biology 13:253–264.

Strayer, D., and J. Ralley. 1991. The freshwater mussels (Mollusca: Unionoidea) of the upper Delaware River basin, New York. American Malacological Bulletin 9:21–25.

Strayer, D. L. 1999a. The statistical power of presence–absence data to detect population declines. Conservation Biology 13:1034–1038.

Strayer, D. L. 1999b. Use of flow refuges by unionoid mussels in rivers. Journal of the North American Benthological Society 18:468–476.

Strayer, D. L., N. F. Caraco, J. J. Cole, S. Findlay, and M. L. Pace. 1999. Transformation of freshwater ecosystems by bivalves: a case study of zebra mussels in the Hudson River. BioScience 49:19–27.

Strayer, D. L., S. Claypool, and S. Sprague. 1997. Assessing unionid populations with quadrats and timed searches. Pages 163–169 *in* K. S. Cummings, A. C. Buchanan, C. A. Mayer, and T. J. Naimo, editors. Conservation and management of freshwater mussels II. Initiatives for the future. Upper Mississippi River Conservation Committee, Rock Island, Illinois.

Strayer, D. L., J. J. Cole, G. E. Likens, and D. C. Buso. 1981. Biomass and annual production of the freshwater mussel *Elliptio complanata* in an oligotrophic softwater lake. Freshwater Biology 11:435–440.

Strayer, D. L., and A. R. Fetterman. 1999. Changes in the distribution of freshwater mussels (Unionidae) in the upper Susquehanna River basin, 1955–1997. American Midland Naturalist 142:328–339.

Strayer, D. L., D. C. Hunter, L. C. Smith, and C. Borg. 1994. Distribution, abundance, and role of freshwater clams (Bivalvia: Unionidae) in the freshwater tidal Hudson River. Freshwater Biology 31:239–248.

Strayer, D. L., and K. J. Jirka. 1997. The pearly mussels of New York State. Memoirs of the New York State Museum 26:1–113.

Strayer, D. L., and L. C. Smith. 1996. Relationships between zebra mussels (*Dreissena polymorpha*) and unionid clams during the early stages of the zebra mussel invasion of the Hudson River. Freshwater Biology 36:771–779.

Strayer, D. L., S. Sprague, and S. Claypool. 1996. A range-wide assessment of populations of *Alasmidonta heterodon*, an endangered freshwater mussel (Bivalvia: Unionidae). Journal of the North American Benthological Society 15:308–317.

Thomas, L., and C. J. Krebs. 1997. A review of statistical power analysis software. Bulletin of the Ecological Society of America 76:126–139.

Thompson, S. K. 1992. Sampling. Wiley, New York.

Thompson, S. K., and G. A. F. Seber. 1996. Adaptive sampling. Wiley, New York.

Turner, M. G., R. H. Gardner, and R. V. O'Neill. 2001. Landscape ecology in theory and practice: pattern and process. Springer-Verlag, New York.

Tyrrell, M., and D. J. Hornbach. 1998. Selective predation by muskrats on freshwater mussels in 2 Minnesota rivers. Journal of the North American Benthological Society 17:301–310.

van der Schalie, H. 1938. The naiad fauna of the Huron River, in southeastern Michigan. Miscellaneous Publications of the Museum of Zoology, University of Michigan 40:1–83.

Vaughn, C. C., and C. C. Hakenkamp. 2001. The functional role of burrowing bivalves in freshwater ecosystems. Freshwater Biology 46:1431–1446.

Vaughn, C. C., and C. M. Taylor. 1999. Impoundments and the decline of freshwater mussels: a case study of an extinction gradient. Conservation Biology 13:912–920.

Vaughn, C. C., C. M. Taylor, and K. J. Eberhard. 1997. A comparison of the effectiveness of timed searches vs. quadrat sampling in mussel surveys. Pages 157–162 *in* K. S. Cummings, A. C. Buchanan, C. A. Mayer, and T. J. Naimo, editors. Conservation and management of freshwater mussels II. Initiatives for the future. Upper Mississippi River Conservation Committee, Rock Island, Illinois.

Venette, R. C., R. D. Moon, and W. D. Hutchison. 2002. Strategies and statistics of sampling for rare individuals. Annual Review of Entomology 47:143–174.

Watters, G. T. 1992. Unionids, fishes, and the species–area curve. Journal of Biogeography 19:481–490.

Watters, G. T. 1994. Sampling freshwater mussel populations: the bias of muskrat middens. Walkerana 7:63–69.

Welker, M., and N. Walz. 1998. Can mussels control the plankton in rivers? A planktological approach applying a Lagrangian sampling strategy. Limnology and Oceanography 43:753–762.

White, G. C., and K. P. Burnham. 1999. Program MARK—survival estimation from populations of marked animals. Bird Study 46(supplement):S120–S139.

Wilcox, D. B., D. D. Anderson, and A. C. Miller. 1993. Survey procedures and decision criteria for estimating the likelihood that *Lampsilis higginsi* is present in areas within the Upper Mississippi River system. Pages 163–169 *in* K. S. Cummings, A. C. Buchanan, and L. M. Koch, editors. Conservation and management of freshwater mussels. Upper Mississippi River Conservation Committee, Rock Island, Illinois.

Williams, B. K., J. D. Nichols, and M. J. Conroy. 2001. Analysis and management of animal populations: modeling, estimation, and decision making. Academic Press, New York.

Williams, J. C., and G. A. Schuster. 1989. Freshwater mussel investigations of the Ohio River, mile 317.0 to 981.0. Kentucky Department of Fish and Wildlife Resources, Frankfort.

Williams, J. D., M. L. Warren, Jr., K. S. Cummings, J. L. Harris, and R. J. Neves. 1993. Conservation status of the freshwater mussels of the United States and Canada. Fisheries 18(9):6–22.

Yamamura, K. 1999. Transformation using (x + 0.5) to stabilize the variance of populations. Researches on Population Ecology 41:229–234.

Yeager, B. L., and R. J. Neves. 1986. Reproductive cycle and fish hosts of the rabbit's foot mussel, Quadrula cylindrical strigillata (Mollusca: Unionidae) in the upper Tennessee River drainage. American Midland Naturalist 116:329–340.

Ziuganov, V., A. Zotin, L. Nezlin, and V. Tretiakov. 1994. The freshwater pearl mussels and their relationships with salmonid fish. VNIRO Publishing House, Moscow.

Index